T0359566

AJ Sadler

Mathematics Applications

Student Book

Unit 4

NELSON
A Cengage Company

Australia • Brazil • Japan • Korea • Mexico • Singapore • Spain • United Kingdom • United States

Mathematics Applications Unit 4
1st revised Edition
A J Sadler

Publishing editor: Robert Yen
Project editor: Alan Stewart
Cover design: Chris Starr (MakeWork)
Text designers: Sarah Anderson, Nicole Melbourne,
Danielle Maccarone
Permissions researcher: Kaitlin Jordan
Answer checker: George Dimitriadis
Production controller: Erin Dowling
Typeset by: Nikki M Group Pty Ltd

Any URLs contained in this publication were checked for
currency during the production process. Note, however, that
the publisher cannot vouch for the ongoing currency of
URLs.

For product information and technology assistance,
in Australia call 1300 790 853;
in New Zealand call 0800 449 725

For permission to use material from this text or product, please email
aust.permissions@cengage.com

National Library of Australia Cataloguing-in-Publication Data
Sadler, A. J., author.
Mathematics applications. Unit 4 / A J Sadler.

1st revised edition
9780170395069 (paperback)
For secondary school age.

Mathematics--Study and teaching (Secondary)--Australia.
Mathematics--Problems, exercises, etc.
Mathematics--Textbooks.

Cengage Learning Australia
Level 7, 80 Dorcas Street
South Melbourne, Victoria Australia 3205

Cengage Learning New Zealand
Unit 4B Rosedale Office Park
331 Rosedale Road, Albany, North Shore 0632, NZ

For learning solutions, visit cengage.com.au

Printed in China by 1010 Printing International Limited
8 9 10 11 25 24

PREFACE

This text targets Unit Four of the West Australian course *Mathematics Applications*, a course that is organised into four units altogether, the first two for year eleven and the last two for year twelve.

This West Australian course, *Mathematics Applications*, is based on the Australian Curriculum Senior Secondary course *General Mathematics*. At the time of writing there are only small differences between the Unit Four content of these two courses, the main difference being that specific examples mentioned in the Australian Curriculum are moved to the glossary of the West Australian syllabus, or removed. Hence this text is also suitable for anyone following Unit Four of the Australian Curriculum course, *General Mathematics*.

The book contains text, examples and exercises containing many carefully graded questions. A student who studies the appropriate text and relevant examples should make good progress with the exercise that follows.

The book commences with a section entitled **Preliminary work**. This section briefly outlines work of particular relevance to this unit that students should either already have some familiarity with from the mathematics studied in earlier years, or for which the brief outline included in the section may be sufficient to bring the understanding of the concept up to the necessary level.

As students progress through the book they will encounter questions involving this Preliminary work in the **Miscellaneous exercises** that feature at the end of each chapter. These miscellaneous exercises also include questions involving work from preceding chapters to encourage the continual revision needed throughout the unit.

Some chapters commence with, or contain, a '**Situation**' or two for students to consider, either individually or as a group. In this way students are encouraged to think and discuss a situation, which they are able to tackle using their existing knowledge, but which acts as a forerunner and stimulus for the ideas that follow. Students should be encouraged to discuss their solutions and answers to these situations and perhaps to present their method of solution to others. For this reason answers to these situations are generally not included in the book.

At times in this series of books I have found it appropriate to go a little beyond the confines of the syllabus for the unit involved. In this regard readers will find that in this book I briefly revise simple interest before proceeding to compound interest and, on the basis that centred moving averages might be assumed content under the general inclusion of moving averages, I include mention of this idea. (However if one considers a centred moving average as a weighted average then the glossary for the course, at the time of writing this text, says that weighted moving averages are beyond the requirements for the unit.) I leave it up to the readers and teachers to decide whether to cover these aspects or not.

Alan Sadler

ISBN 9780170395069

CONTENTS

① TIME SERIES DATA   2

② MOVING AVERAGES AND SEASONAL EFFECTS   18

③ FINANCE I: SAVING AND BORROWING   52

IMPORTANT N⬤TE

This series of texts has been written based on my interpretation of the appropriate *Mathematics Applications* syllabus documents as they stand at the time of writing. It is likely that as time progresses some points of interpretation will become clarified and perhaps even some changes could be made to the original syllabus. I urge teachers of the *Mathematics Applications* course, and students following the course, to check with the appropriate curriculum authority to make themselves aware of the latest version of the syllabus current at the time they are studying the course.

Acknowledgements

As with all of my previous books I am again indebted to my wife, Rosemary, for her assistance, encouragement and help at every stage.

To my three beautiful daughters, Rosalyn, Jennifer and Donelle, thank you for the continued understanding you show when I am 'still doing sums' and for the love and belief you show.

I thank my good friend and ex-colleague Theo Wieman for his much appreciated wise counsel.

To the delightfully supportive team at Cengage – I thank you all.

Alan Sadler

PRELIMINARY WORK

This book assumes that you are already familiar with a number of mathematical ideas from your mathematical studies in earlier years.

This section outlines the ideas which are of particular relevance to Unit Four of the *Mathematics Applications* course and for which some familiarity will be assumed, or for which the brief explanation given here may be sufficient to bring your understanding of the concept up to the necessary level.

Read this 'preliminary work' section and if anything is not familiar to you, and you don't understand the brief mention or explanation given here, you may need to do some further reading to bring your understanding of those concepts up to an appropriate level for this unit. (If you do understand the work but feel somewhat 'rusty' with regards to applying the ideas some of the chapters afford further opportunities for revision as do some of the questions in the miscellaneous exercises at the end of chapters.)

- Chapters in this book will continue some of the topics from this Preliminary work by building on the assumed familiarity with the work.

- The miscellaneous exercises that feature at the end of each chapter may include questions requiring an understanding of the topics briefly explained here.

Number

The understanding and appropriate use of the rule of order, fractions, decimals, percentages, rounding, truncation, square roots and cube roots, numbers expressed with positive integer powers, e.g. $2^3, 5^2, 2^5$, expressing numbers in standard form, e.g. $2.3 \times 10^4 (= 23\,000)$, $5.43 \times 10^{-7} (= 0.000\,000\,543)$, also called scientific notation, and familiarity with the symbols $>, \geq, <$ and $\leq$ is assumed.

Percentages

It is assumed that you are able to express one quantity as a percentage of another quantity and find a percentage of a quantity. It is also assumed that you are able to increase or decrease an amount by a certain percentage by appropriate multiplication. For example:

- To increase an amount by 25% we multiply by 1.25.
- To decrease an amount by 5% we multiply by 0.95.

Typical percentage questions:

I   Express 28 out of 40 as a percentage.

First express as a fraction: $\dfrac{28}{40}$

Then multiply by 100: $\dfrac{28}{40} \times 100 = 70\%$

Thus 28 out of 40 is 70%.

II   Find 28% of $40.

Either: Divide by 100 to find 1%: $\dfrac{\$40}{100}$

Then multiply by 28 to find 28%: $\dfrac{\$40}{100} \times 28$

= $11.20

Or: Use the decimal equivalent of 28 %

28% of $40   = $40 × 0.28
             = $11.20

The following examples of 'typical percentage questions' will only show the 'decimal equivalent' method.

III   Increase $40 by 28%.

$40 increased by 28%   = $40 × 1.28
                       = $51.20

IV   Decrease $40 by 28%

$40 decreased by 28%   = $40 × 0.72
                       = $28.80

IV   28% of an amount is $40. Find the amount.

28% of an amount is $40

100% of the amount   = $\dfrac{\$40}{0.28}$

= $142.86 (nearest cent).

Straight line graphs

If points lie in a straight line their coordinates, (x, y), obey a rule of the form

$$y = mx + c$$

and the relationship between x and y is said to be *linear*.

In this rule m is the gradient of the straight line and the point $(0, c)$ is where the line cuts the vertical axis.

Hence the line $y = 2x + 1$ has a gradient of 2 and cuts the y-axis at $(0, 1)$.

The line $y = -x + 5$ has gradient of -1 and cuts the y-axis at $(0, 5)$.

In the equation $y = mx + c$ the value of m, the gradient, tells us the amount by which y increases for each unit increase in x.

Solving equations

It is assumed that you are already familiar with the idea that solving an equation involves finding the value(s) the unknown can take that make the equation true.

For example, $x = 5.5$ is the solution to the equation

$$15 - 2x = 4$$

because

$$15 - 2(5.5) = 4.$$

To solve an equation, e.g. $15 - 2x = 4$, we could proceed *mentally*:

We know that fifteen take *eleven* equals four.
Thus $2x = 11$ and so $x = 5.5$.

We could use the *solve facility* of some calculators.

Or we could use a *step by step approach* to isolate x:

Given the equation:	$15 - 2x$	$=$	4
Add $2x$ to both sides to make the x term positive:	15	$=$	$4 + 2x$
Subtract 4 from both sides to isolate $2x$:	$15 - 4$	$=$	$2x$
$\therefore$	11	$=$	$2x$
Divide both sides by 2 to isolate x:	5.5	$=$	x
$\therefore$	x	$=$	5.5

It is also anticipated that you are able to solve more involved linear equations, for example:

$$5x - 7 - 2x - 8 = 30 - 6x \qquad \text{Solution:} \quad x = 5$$

$$\frac{x + 3}{2} - 1 = 5 \qquad \text{Solution:} \quad x = 9$$

$$2(x + 3) - 3(2x + 1) = -5 \qquad \text{Solution:} \quad x = 2$$

and simultaneous linear equations, for example:

$$\begin{cases} 3x + 2y = 11 \\ x + 2y = 1 \end{cases} \qquad \begin{cases} 5x - 2y = 6 \\ 3x + 2y = 26 \end{cases} \qquad \begin{cases} 2x + 3y = 12 \\ x + 4y = 11 \end{cases}$$

Solution: $\qquad x = 5, y = -2 \qquad\qquad x = 4, y = 7 \qquad\qquad x = 3, y = 2$

In this unit we may also encounter equations involving exponents (powers), for example

$$x^4 = 81 \qquad 2^x = 64 \qquad 2^x = 11$$

Solve these mentally if the numbers involved allow it, otherwise use the solve facility of some calculators.

$$x^4 = 81 \qquad\qquad 2^x = 64 \qquad\qquad 2^x = 11$$
$$x = 3 \qquad\qquad\; x = 6$$

solve($2^x = 11, x$)

$\{x = 3.459431619\}$

$$x = 3.46 \text{ (correct to 2 decimal places)}$$

ISBN 9780170395069

Simple and compound interest

Suppose we were to invest $500 for 3 years in an account paying interest at the rate of 10% per annum.

Considering a **simple interest** approach in which the same amount of interest is added each year:

$$\begin{aligned} \text{Value after 3 years} &= \$500 + 3 \times 10\% \text{ of } \$500 \\ &= \$500 + 3 \times \$50 \\ &= \$650 \end{aligned}$$

Alternatively, considering **compound interest**, in which the interest earned in one year itself earns interest in the next year:

$$\begin{aligned} \text{Value after 1 year} &= \$500 + 10\% \text{ of } \$500 \\ &= \$500 \times 1.1 \\ &= \$550 \end{aligned}$$

$$\begin{aligned} \text{Value after 2 years} &= \$550 \times 1.1 \quad (\text{i.e. } \$500 \times 1.1^2) \\ &= \$605 \end{aligned}$$

$$\begin{aligned} \text{Value after 3 years} &= \$500 \times 1.1^3 \\ &= \$665.50 \end{aligned}$$

If we want the value after n years:

$$\text{Value after } n \text{ years} = \text{Initial value} \times 1.1^n$$

Recursion

It is assumed that you are familiar with the idea of defining a sequence by a **recursive rule** or **recurrence relationship**, i.e. a rule which tells us how the terms of the sequence recur. For example, the following rule

$$T_1 = 5, \qquad T_{n+1} = T_n + 3 \qquad (\text{or } T_1 = 5, T_n = T_{n-1} + 3)$$

tells us that the first term of the sequence is 5 and each term after that is obtained by adding 3 to the previous term

Hence the sequence is: 5,   8,   11,   14,   17,   20,   23,   …

With adjacent terms having a common difference this is an example of an **arithmetic sequence**, also called an **arithmetic progression** or **AP**.

Thus all arithmetic sequences are of the form:

$$a, \qquad a+d, \qquad a+2d, \qquad a+3d, \qquad a+4d, \qquad a+5d, \qquad a+6d, \qquad …$$

In this general form we have a first term of 'a' and **common difference** 'd'.

The nth term is then given by $T_n = a + (n-1)d$.

Sequences which progress such that each term of the sequence is the previous term multiplied by some constant number are said to be **geometric sequences**, **geometric progressions**, or **GPs**.

Thus all GPs are of the form:

$$a, \qquad ar, \qquad ar^2, \qquad ar^3, \qquad ar^4, \qquad ar^5, \qquad ar^6, \qquad …$$

In this general form we have a first term of 'a' and **common ratio** 'r'.

The nth term is then given by $T_n = a \times r^{n-1}$.

GPs will have a recursive rule of the form $T_1 = a$, $T_{n+1} = r \times T_n$.

It is also assumed that you are familiar with recurrence rules of the form:

$$T_1 = a, \qquad\qquad T_{n+1} = b \times T_n + c.$$

This type of recursive rule will generate sequences for which each term, after the first, is obtained from the previous term by multiplying by some constant value, b, and then adding a constant value, c.

For example $\quad T_1 = 4000, \qquad\qquad T_{n+1} = 1.01 \times T_n - 100.$

Thus $\qquad\quad T_1 = 4000, \qquad\qquad$
$$\begin{aligned} T_2 &= 1.01 \times T_1 - 100 \\ &= 1.01 \times 4000 - 100 \\ &= 3940 \end{aligned}$$

$$\begin{aligned} T_3 &= 1.01 \times T_2 - 100 \\ &= 1.01 \times 3940 - 100 \\ &= 3879.4 \end{aligned}$$

$$\begin{aligned} T_4 &= 1.01 \times T_3 - 100 \\ &= 1.01 \times 3879.4 - 100 \\ &= 3818.194 \qquad\qquad \text{etc.} \end{aligned}$$

Recurrence relations of this form are particularly relevant to this unit.

Graphs or networks

In mathematics, a network or graph is a set of points, called **vertices**, connected by a set of lines, called **edges**.

- The junctions and endpoints are called **vertices** (singular: **vertex**).

4 vertices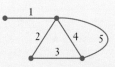

- The connections are called **edges**.

5 edges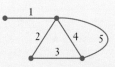

Graphs may be directed or undirected.

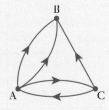

A directed graph
(Also referred to as a digraph)
The **directed edges** are sometimes
referred to as **arcs**.

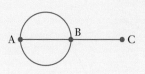

An undirected graph

ISBN 9780170395069

Some graphs show 'weights' on each edge, perhaps distance, cost, time etc.

Such graphs are referred to as weighted graphs. For example:

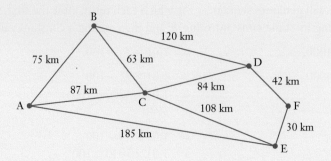

Whilst it is anticipated that from your study of Unit Three of *Mathematics Applications* you are familiar with many of the terms associated with graph theory (e.g. multiple edges, adjacent vertices, simple graphs, connected graphs, planar graphs, subgraphs, bridges, walks, paths, cycles, etc.), in this unit we will be mainly focusing on analysing weighted graphs and in one such aspect an understanding of the word **tree** in the context of graph theory is important, and in another aspect **bipartite** graphs are significant. These two terms in particular are further explained below:

Any connected simple graph that contains no cycles is called a tree:

The following graphs are trees:

 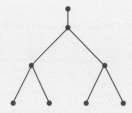 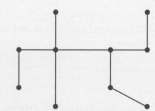

The following are not trees:

 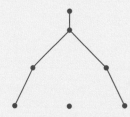 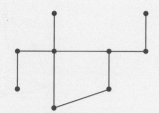

Contains a cycle. Not connected. Contains a cycle.
Therefore not a tree. Therefore not a tree. Therefore not a tree.

ISBN 9780170395069

Consider the following graph:

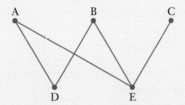

Notice that if we consider the five vertices A, B, C, D and E as being in two sets:

{A, B, C} and {D, E}

then every one of the five edges joins a vertex from one set to a vertex of the other set.

AD, AE, BD, BE, CE.

A graph in which the vertices can be split into two groups such that every edge joins a vertex from one group to a vertex of the other group, i.e. no edges join vertices from the same group, is called a **bipartite** graph. Thus, in a bipartite graph, every pair of adjacent vertices involves one vertex from one group and the other vertex from the other group.

For example, the bipartite graph below shows which of the three areas, midwifery (Mid), accident and emergency (A & E), and the intensive care unit (ICU), three nurses are qualified to work in.

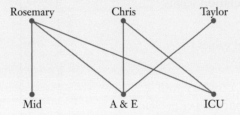

Bivariate data

If a situation involves two data sets, for example the heights and weights of a number of people, we have **bivariate data**. We might then be interested in investigating whether there is an **association** between the data sets, for example, is there some association between a child's age and their foot size? Do older children generally have bigger feet? Does a child's age in any way explain their foot size? In this case, age would be the **explanatory variable**, also called the independent or predictor variable. Foot size would be the **response variable**, also called the dependent or predicted variable.

Shutterstock.com/Beer Pintusan

ISBN 9780170395069

Regression

The scattergraph on the right shows the consumption of a particular food commodity on the horizontal x-axis, and heart disease death rate on the vertical y-axis, for twenty countries.

The location of the points suggests that as one of the measures increases then the other does too.

This apparent **correlation** between the two variables could be modelled by drawing a **line of best fit** or **trend line**, as shown on the next graph.

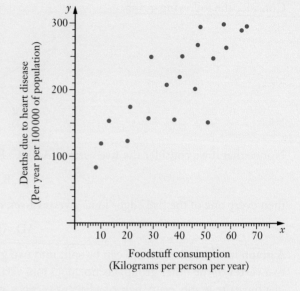

Foodstuff consumption
(Kilograms per person per year)

With one variable generally increasing as the other increases this line of best fit has a positive gradient.

We say that a **positive linear correlation** seems to exist.

The line of best fit allows predicted values to be suggested.

For example we could suggest that a country with an intake of the foodstuff of 45 kilograms per person per year would have a heart disease death rate of approximately 230 persons per year per 100 000 of population.

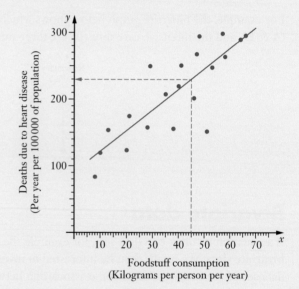

Foodstuff consumption
(Kilograms per person per year)

Rather than having to draw this 'line of best fit' by eye we can obtain its equation and predict values using the **linear regression** techniques of some calculators and spreadsheets.

The display on the right gives the equation of the line of best fit, or **least squares regression line**, as

$$y = 3.10x + 90.9$$

> Linear Reg
> $a = 3.09976491$
> $b = 90.894086$
> $r = 0.82280944$
> $r^2 = 0.67701539$
> $y = ax + b$

ISBN 9780170395069

Given that the data for the previous scattergraph is as shown below **make sure that you can obtain this equation from your calculator**.

x	57	20	13	38	51	48	58	35	29	8	28	21	10	40	41	46	53	64	66	47
y	299	124	154	156	152	295	264	208	250	84	158	175	120	220	251	202	248	290	296	268

We can then calculate (or determine automatically from our calculator) a predicted value for 'y' (in this case deaths per year per 100 000 of population) for an 'x' value (in this case foodstuff consumption) of 45.

$$y = 3.0998(45) + 90.8941 \text{ i.e. approximately } 230.$$

To emphasise the predicted nature of y this is sometimes written as $\hat{y}$, pronounced 'y hat'. Thus for $x = 45$, $\hat{y} \approx 230$.

The 'r' given in the previous calculator display is called **Pearson's correlation coefficient**. It will lie in the interval $-1 \le r \le +1$. This correlation coefficient is a measure of the strength and nature of the linear association, or correlation, between the variables. It informs us how closely the relationship between two sets of data approximates to a linear relationship.

A correlation coefficient of -1 indicates a perfect linear relationship with a negative gradient (as one variable increases the other decreases) and a correlation coefficient of $+1$ indicates a perfect linear relationship with a positive gradient (as one variable increases the other increases).

(The display also shows r^2, the **coefficient of determination**. This gives the proportion of the variation that can be explained by the linear relationship. I.e. in the previous situation the r^2 value of 0.677 means that approximately 68% of the variation in death rates due to heart disease can be explained by, or accounted for, by the variation in the foodstuff consumption.)

If there is an association between the two variables then as one changes we should expect the other to change, but that is *not* to say that the changes in one variable *cause* the changes in the other. There could be other factors at work.

If we use the fact that a strong correlation seems to exist, to predict a 'y-value' for some 'x-value', we can be more confident of this prediction if **interpolation** is involved, i.e. if the x-value used is within the range of x-values covered by the given data.

If instead we are **extrapolating**, i.e. using an x-value outside the range covered by the x-values in the given data, we would be less confident with any prediction made, our confidence decreasing the further the x-value is from the range of x-values in the given data.

Use of technology

You are encouraged to use

- calculators to solve equations when necessary, to display the terms of a recursively defined sequence when appropriate, and to display graphs,
- spreadsheets on a computer (or calculator),
- the internet for research.

$$\begin{cases} x + y = 1200 \\ 7x + 10y = 9150 \end{cases} \bigg| \; x, y$$
$$\{x = 950, y = 250\}$$

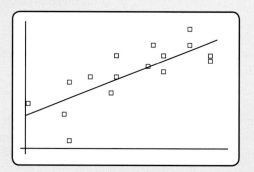

	A	B	C	D	E	F	G
1	Physics score	Maths score					
2	6	19					
3	9	12					
4	11	27					
5	12	36					
6	14	42					
7	14	34					
8	16	32					
9	16	27					
10	17	44					
11	19	41					
12							
13							

Physics and Mathematics scores

$y = 1.9829x + 4.8291$

Score in Mathematics vs Score in Physics

Shutterstock.com/Sergey Nivens

ISBN 9780170395069

Calculators and computers can be particularly useful if we wish to display the terms of sequences. Consider for example the growth in the value of a house that is initially valued at $500 000 and is subject to an annual increase in value of 6.4%.

$$
\begin{array}{llll}
\text{Initial value} & = & \$500\,000 & \leftarrow T_1 \\
\text{Value after 1 year} & = & \$500\,000 \times 1.064 & \leftarrow T_2 \\
\text{Value after 2 years} & = & \$500\,000 \times 1.064^2 & \leftarrow T_3 \\
\text{Value after 3 years} & = & \$500\,000 \times 1.064^3 & \leftarrow T_4 \quad \text{etc.}
\end{array}
$$

These values form a geometric sequence with
$$T_{n+1} = T_n \times 1.064$$
and
$$T_1 = 500\,000$$

We could display the terms of our sequence on a calculator or spreadsheet and view the progressive year by year values, as shown below.

	A	B	C	D
1	Initial value			$500,000.00
2	Percentage increase			6.40
3	Value at end of year		1	$532,000.00
4			2	$566,048.00
5			3	$602,275.07
6			4	$640,820.68
7			5	$681,833.20
8			6	$725,470.52
9			7	$771,900.64
10			8	$821,302.28
11			9	$873,865.63
12			10	$929,793.03
13			11	$989,299.78
14			12	$1,052,614.96

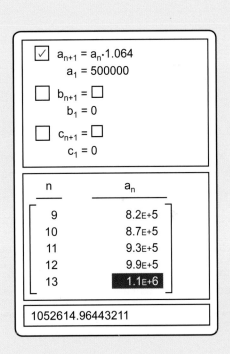

These tables allow us to see, for example, that the value of the house will reach one million dollars shortly after the end of the eleventh year, i.e. early in the 12th year.

However note carefully that in this situation, with the recursive definition

$$T_{n+1} = T_n \times 1.064 \qquad \text{and} \qquad T_1 = 500\,000$$

T_1 is the value after **zero** years. Hence we must remember that if we use the ability of a calculator to generate the terms of the sequence, according to the recursive rule given above, then the value after n years will be given by T_{n+1}. I.e. in the calculator display above, $n = 13$ gives the value at the end of 12 years.

One way to avoid this possible source of confusion would be to use the ability of some calculators to accept a sequence defined using T_0 as the 1st term.

I.e. define the sequence as:

$$T_{n+1} = T_n \times 1.064 \quad \text{and} \quad T_0 = 500\ 000,$$

as shown on the right.

Under such a definition T_n would indeed be the value after n years.

Alternatively we could use

$$T_{n+1} = T_n \times 1.064 \quad \text{and} \quad T_1 = 500\ 000 \times 1.064$$

and again T_n would be the value after n years.

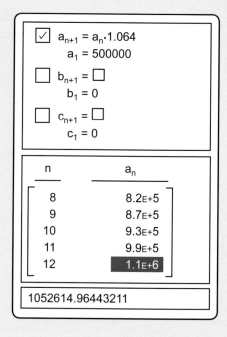

The statistical investigation process

Given some real world problem, for example, estimating the future health needs for a particular region with regards to number of hospital beds required, the statistical investigation process follows the following steps:

(1) Clarify the problem and formulate one or more questions that can be answered with data.

(2) Design and implement a plan to collect and obtain appropriate data.

(3) Select and apply appropriate graphical or numerical techniques to analyse the data.

(4) Interpret the results of this analysis and relate the interpretation to the original question; communicate findings in a systematic and concise manner.

(As stated in the Australian Curriculum Glossary for General Mathematics and the West Australian Glossary for Mathematics Applications.)

Algorithms

Remember, an algorithm is a defined set of steps that that can be applied and systematically followed to solve a particular type of problem. In Unit Three of *Mathematics Applications* we encountered an algorithm for determining the shortest path between two vertices in a network.

ISBN 9780170395069

1.

Time series data

- Time series data
- Making predictions
- Miscellaneous exercise one

Situation One

In the year 2015 Mrs Maggie Smithson sells a valuable painting at auction for $1 243 000.

Mrs Smithson's accountant informs her that for taxation purposes he needs to know what the value of the painting was when it first came into her possession.

Mrs Smithson says that it was left to her in 2001 upon the death of the previous owner, Mrs Smithson's aunt, but she did not know the value of the painting at that time. What she did have though was a number of valuations, carried out at various times for insurance purposes.

The accountant listed these insurance valuations, together with the price achieved at auction as follows:

Year of valuation	1948	1968	1988	2008	2015
Valuation	$85 000	$280 000	$345 000	$810 000	$1 243 000

Suggest what you consider to be a reasonable valuation for the painting for the year that it first became the property of Mrs Smithson and explain how you arrived at your estimate and why you think it reasonable.

Situation Two

A group is asked to write a report outlining the future health needs for a particular region. One of the things the group needs is an estimate of the number of hospital beds the region is likely to need in 5, 10 and 15 years time.

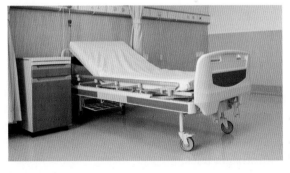

The number of hospital beds available in the region over the last twenty years, recorded every 2 years, is known and is shown in the following table:

Time	20 years ago	18 years ago	16 years ago	14 years ago	12 years ago	10 years ago
Number of hospital beds	3270	3350	3510	3520	3760	3900

Time	8 years ago	6 years ago	4 years ago	2 years ago	Now	
Number of hospital beds	4100	4420	4790	5140	5600	

Assuming the above numbers did just about meet the needs of the community at the time, use the figures to make predictions for the number of hospital beds this region is likely to require in 5, 10 and 15 years' time and explain how you arrived at your predictions.

The situations of the previous page each involve measurements of something where the time the measurement is made is also recorded. In each situation the given information is an example of **time series** data.

Each of the situations involved the statistical investigation process:

(1)   Clarify the problem and formulate appropriate questions.

In Situation One we needed to determine the value of the painting when it came into Mrs Smithson's possession. Not knowing this information we needed to ask what was known about its value at other times.

In Situation Two we needed to determine the number of hospital beds required for this region in 5, 10 and 15 years' time. To determine this we needed to ask for historical data about the number of beds required in the region.

(2)   Plan the collection and use of appropriate data.

In Situation One the plan was to use previous insurance valuations we were given.

In Situation Two the plan was to use previous data that was available regarding bed numbers.

(3)   Apply appropriate techniques to analyse the data.

This was your task.

Did you consider graphing the information given in each of the situations?

Graphing time series data can allow underlying **trends** to be seen and predictions to be made. (The trend is the general direction of the time series, e.g. increasing, decreasing, etc.)

(4)   Interpret the results in order to answer the original question and communicate your findings.

This also was your task.

If you did graph the data of the two situations you may have come to the conclusion that attempting to fit a straight line to the data might not be a suitable choice of 'model'. Perhaps you tried another model or, with each set of points perhaps suggesting a suitable curve that could be drawn 'freehand' maybe you drew such a curve and read off the required values, or maybe you simply joined adjacent points with straight lines.

One of the situations required us to **extrapolate** beyond the known values so predictions obtained in this way would have to be viewed with caution. However, though extrapolation can be unreliable, it is sometimes the best we can do. With the situation that did not require extrapolation, but instead required us to determine a value between known points (**interpolation**), our prediction should be more reliable. However, even with interpolation, out-of-the-ordinary events and short-term fluctuations can sometimes give values that are not in line with normal trend. For example we might expect monthly visitor numbers to a city to peak higher than a general trend would suggest if the city were to host an Olympic Games for a month. We might expect daily traffic flow along a particular major road to drop below the figure a general trend would suggest if there were major roadworks taking place along the road and diversions were in place, etc.

Time series data

The diagram on the right shows a **line graph**, i.e. a graph in which each point is joined to the next by a straight line.

This particular line graph involves observations that are ordered according to time, in this case in five-year intervals from 1970 to 2015. The graph involves **time series data**.

Note that when graphing time series data, time is plotted on the horizontal axis.

The joining of the points allows the underlying trend to be seen.

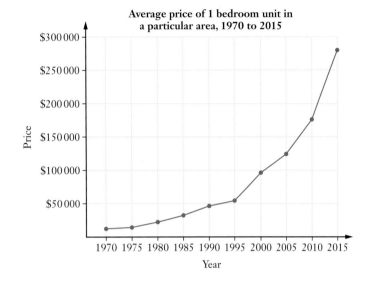

The graph shows an **increasing trend** – the prices increase with the passing of time.

Drawing straight lines between adjacent points can allow underlying patterns to become more evident. In the graph shown below there is clearly an increasing trend but can you also see the five point repeating pattern?

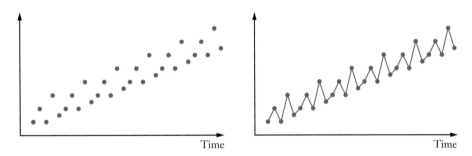

We will consider repeating patterns like this in more detail in the next chapter.

It can sometimes be the case that time series data is collected at intervals that are not all of the same length (as was the case in Situation One). Consider the following situation for example:

An environmental group, concerned with the levels of a particular pollutant in a river, collected and analysed samples from the river over a period of some years. The results are shown in the table below.

Month and year	Jan 2010	Jul 2010	Jan 2011	Jul 2011	Jan 2012	Jul 2012	Jan 2013	Apr 2013	Jul 2013	Oct 2013	Jan 2014	Apr 2014	Jul 2014
Pollutant level (units/m³)	2.04	2.09	2.24	2.51	2.44	2.76	2.84	2.92	3.06	3.04	3.14	3.21	3.37

Notice that in 2010, 2011 and 2012 levels were recorded half-yearly, but in 2013 levels were recorded quarterly. The graph of this data is shown on the next page.

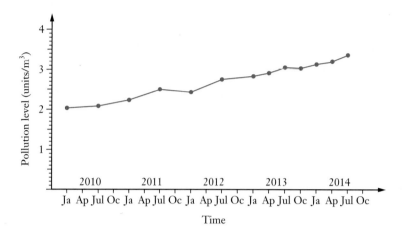

- Notice that the horizontal axis uses a consistent time scale throughout. The fact that from January 2013 onwards pollutant levels are given 4 times per year, rather than the 2 per year prior to this, changes the number of points plotted per year not the scale on the axis.

Making predictions

If we want to analyse the data mathematically we might need to number the dates involved, rather than to attempt putting Jan 2010, Jul 2010 etc. into our calculator.

Particular care needs to be taken if the frequency of data collection changes as in this pollution levels example. Noticing that from 2013 onwards pollutant levels are given 4 times per year we could number the data points as follows:

Data point (n)	1	3	5	7	9	11	13	14	15	16	17	18	19
Month and year	Jan 2010	Jul 2010	Jan 2011	Jul 2011	Jan 2012	Jul 2012	Jan 2013	Apr 2013	Jul 2013	Oct 2013	Jan 2014	Apr 2014	Jul 2014
Pollution (P) (units/m^3)	2.04	2.09	2.24	2.51	2.44	2.76	2.84	2.92	3.06	3.04	3.14	3.21	3.37

The graph would then be as shown below:

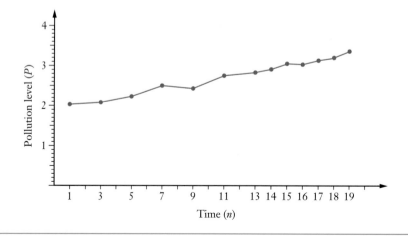

With the data following a reasonably straight line we could then use our knowledge of linear regression to determine the least squares regression line:

$$P = 0.0735n + 1.91$$

(and a correlation coefficient of 0.99).

Asked to predict pollution levels for January 2018 we note that this would be for when n has the value 33.

For this value of n the regression line gives a predicted value of 4.3 for P.

Thus the predicted pollution level for January 2018 is 4.3 units/m^3.

However, even with our 'eye-balling' of the points as being reasonably linear, and the correlation coefficient of 0.99 indicating that using a linear model is appropriate, January 2018 is well beyond dates during which pollution readings were collected so the reliability of the prediction is extremely questionable. If the trend continues it may be a reasonable prediction but from July 2014 to Jan 2018 many things could occur to alter the trend.

Of course in some cases linear regression may not be suitable – as in the two situations at the beginning of this chapter. In such cases other models available on some calculators could be used, and you are encouraged to explore such possibilities, or simply 'eye-ball' the data and continue the trend as seems appropriate. Whilst in this text we will concentrate on using a least squares *linear* regression model it is clearly important to be able to recognise situations for which linear regression is not suitable.

Consider again the earlier situation involving the price of a 1 bedroom unit in a particular area. In this case linear regression would be inappropriate because a straight line model would not fit the data points at all well.

If asked to use the graph to predict what the average price was likely to have been in 2008 we could draw the line up from 2008 until it meets our graph and then draw across, as shown on the right, to obtain an estimate, in this case approximately $156000.

Similarly, if asked to estimate the likely average price for the unit in the year 2020 we could continue the trend of the graph, as best we can, and read off a value, in this case approximately $480000.

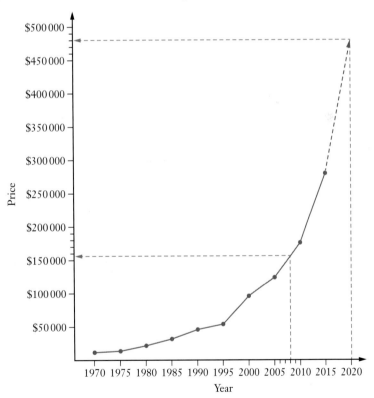

However, as we know, this latter situation again involves extrapolation so any prediction could not be considered to be particularly reliable.

The table below shows the percentage of the Australian population aged 80 or over, for the years 1995 to 2007.

Percentage of the Australian population aged 80 and over

Year	1995	1996	1997	1998	1999	2000	2001	2002	2003	2004	2005	2006	2007
Percentage	2.6	2.6	2.7	2.8	2.8	2.9	3.1	3.2	3.3	3.3	3.4	3.5	3.6

[Source of data: Australian Bureau of Statistics.]

Display this information as a line graph and comment on any trends shown.

Use your graph to suggest what percentage of the Australian population will be aged 80 or over in the year 2020 and in the year 2050, each time discussing the likelihood of your prediction being reliable.

Solution

The graph is shown to the right.

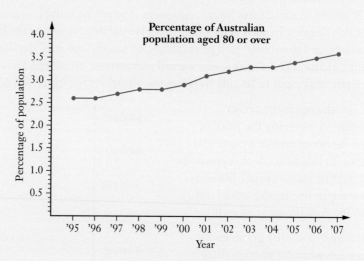

The percentage of the Australian population aged 80 or over shows an increasing trend.

For the period the data describes, i.e. 1995 to 2007, the percentage increases fairly steadily at a rate of approximately 0.1% of the Australian population per annum. (Figures show increase of 1% of population in 12 years.)

Continuing the trend forward to 2020, just over another 12 years, would suggest approximately 4.6% or 4.7% of the population would be 80 or over by then. With the data showing a reasonably linear trend we could use linear regression to obtain a 2020 figure of approximately 4.7%.

This prediction involves extrapolation so the result should not be considered particularly reliable but considering the steadiness of the increase from 1995 to 2007 the estimate may prove to be quite good.

The reasonably steady increase of about 1% every 12 years shown in the graph suggest approximately 7.2% of the population would be 80 or over by 2050. (Linear regression gives 7.4%.) However this requires us to extrapolate so far beyond the known points the 'prediction' is perhaps more an 'educated guess' than a reliable prediction.

→   Try to obtain more up-to-date percentages and see how they compare.   ←

Note:
- Time series data can often be *seasonal* in nature. For example, in the pollution levels in a river situation encountered earlier, the higher temperatures in summer may alter the levels of some pollutants. Thus in addition to showing some overall trend this sort of data may also show variation dependent upon the *season* when the data is collected. We will consider the seasonal nature of some time series data in the next chapter.

- The vertical axis should start from zero if possible. Sometimes this may be inconvenient or impractical and in such cases a break in the axis should be clearly shown (see the diagram on the right).

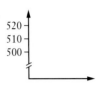

- Remember that time is plotted on the horizontal axis and that you should use the same time scale along the entire axis.

- If you have access to a computer with spreadsheet capability explore its ability, and the ability of your calculator, to draw line graphs and use them to draw some of the graphs requested in the next exercise.

- Remember that predicting into the future can be hazardous. Unanticipated events can occur. Consider for example the graph below, which shows the Dow Jones Index (a measure of the share values of a number of large companies in the USA) for the beginning of March 2003 to the beginning of March 2008.

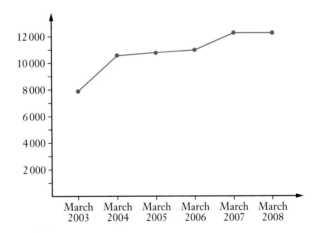

These figures might lead us to expect that by the beginning of March 2009 the index would be perhaps as high as 12 500. In fact it had fallen to just a little over 7000.

Events causing results that seem out of line with a suggested trend may not always be unexpected. The graph on the next page shows the total expenditure, in billions of pounds, of the people visiting London, from elsewhere in the UK, for the years 2008 to 2012.

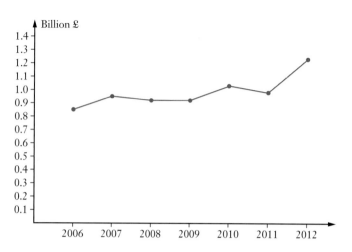

[Source of data: London tourism report 2012/13. London & Partners.]

Does the somewhat higher than trend figure for 2012 indicate increasingly higher figures for 2013, 2014 etc. are likely, or should it be expected with 2012 being the year of the London Olympics?

Exercise 1A

1 Ageing population

The table below shows some real and some predicted figures for the number of persons aged over 65, per 100 persons of working age, in Australia for 1971, 1981, 1991, 2001, 2011, 2021 and 2031.

Number of persons over 65 per 100 persons of working age

Year	1971	1981	1991	2001	2011	2021	2031
Number	13	15	17	18	20	26	31

[Source of data: HBF.]

a We would of course expect the number of people over 65 to increase as the population of Australia increases. Can the increase in the numbers shown in the bottom row of the table (i.e. 13, 15, 17, 18, ...) be attributed to this natural increase in the population or not? Explain your answer.

b Display the data as a line graph.

c Describe any trends shown.

iStock.com/monkeybusinessimages

2 Population of Western Australia

The following table gives the population (to the nearest 1000) of Western Australia in various years.

Year	1915	1925	1935	1945	1955
Population	317 000	378 000	450 000	490 000	669 000

Year	1965	1975	1985	1995	2005
Population	838 000	1 167 000	1 437 000	1 749 000	2 037 000

[Based on Australian Bureau of Statistics data.]

Western Australia

a Display this information as a line graph.

b Comment on any trends shown.

c By 'eye-balling' your graph, predict what the population was in the year 1950 and comment on the likely reliability of your prediction.

d By 'eye-balling' your graph, predict what the population will be in the year 2025 and comment on the likely reliability of your prediction.

e Find out the latest population figure for Western Australia and see how it compares to that predicted by the above figures.

3 Visitors to Australia

The table below shows the number of overseas visitors arriving in Australia for a stay of less than 12 months (called *short-term visitors*) for various years from 1992 to 2013.

Year	1992	1994	1997	2001	2003	2005
Visitors	2 603 100	3 361 600	4 318 000	4 855 800	4 745 800	5 463 000

Year	2007	2009	2010	2011	2012	2013
Visitors	5 588 800	5 490 200	5 790 100	5 770 800	6 032 200	6 380 500

[Source of data: Australian Bureau of Statistics.]

a Display this information as a line graph and comment on any trends shown.

b By 'eye-balling' your graph, predict the number of short-term visitors to Australia in 1999 and in 2020, in each case commenting on the likely reliability of your prediction.

Australia

4 Age at first marriage

In more recent times is the age at which people get married for the first time more or less than it used to be in 'the old days'?

Use the data shown below to formulate a response to the above question, including in your response a graphical display of the data, and commenting on, and describing, any trends shown.

Median age at first marriage in Australia

Year	1966	1970	1974	1980	1982	1984	1985	1988	1990	1991	1992
Men	23.8	23.4	23.3	24.2	24.6	25.1	25.4	26.1	26.5	26.7	26.9
Women	21.2	21.1	20.9	21.9	22.4	22.9	23.2	24.0	24.3	24.5	24.7

Year	1993	1996	1997	2000	2001	2002	2008	2009	2010	2011	2012
Men	27.0	27.6	27.8	28.5	28.7	29.0	29.6	29.6	29.6	29.7	29.8
Women	24.8	25.7	25.9	26.7	26.9	27.1	27.7	27.7	27.9	28.0	28.1

[Source of data: Australian Bureau of Statistics.]

5 Number of marriages

Let us suppose that in a particular country, the number of marriages occurring in various years from 1910 to 2010 were as shown in the following table.

Number of marriages occurring in the country in various years from 1910–2010

Year (Y)	1910	1925	1930	1938	1947	1962	1975	1986	1997	2004	2010
Number (N)	25124	34728	37894	38247	46221	49241	59258	66241	68256	71567	76278

a View the graph of this time series data on a calculator or computer and confirm that the points are suitable for linear modelling.

b With N representing the number of marriages in a year, and Y representing the year, obtain the least squares regression line, $N = mY + b$, for this data, giving m to the nearest integer and b to the nearest 1000.

6 Expanding business

The rental property manager of a real estate business researched data regarding the number of properties managed by the business at various times from 2010 to 2015. The table below shows the data she collected.

Month and year	Mar 2010	Mar 2011	Mar 2012	Mar 2013	Mar 2014	Jun 2014	Sept 2014	Dec 2014	Mar 2015	Jun 2015
No. of properties managed	302	321	408	467	490	492	508	502	526	540

a Commencing with March 2010 as $t = 1$, June 2010 as $t = 2$ etc. determine the equation of the least squares linear regression line for predicting the number of properties managed for a given value of t.

b Assuming this trend continues predict the number of properties managed by the group in December 2017.

7 Caesarean sections

Let us suppose that in a particular region the number of deliveries for all births, over a sixteen-year period, and the number of those deliveries that were performed by caesarean section, were as given in the following table:

Year	1	2	3	4	5	6	7	8
Total births	15 750	16 170	15 971	16 227	16 451	16 332	16 815	16 213
Caesarean	2835	3017	2925	2973	3068	3612	3821	3659

Year	9	10	11	12	13	14	15	16
Total births	16 629	16 873	17 105	16 974	17 043	16 142	17 123	17 260
Caesarean	4083	4431	4862	4904	5082	4758	5161	5178

In year 1 the rate of caesarean deliveries was 2835 out of 15 750, i.e. 18%.

In year 16 the rate of caesarean deliveries was 5178 out of 17 260, i.e. 30%.

If this increase in the caesarean rate over this time were perfectly linear the caesarean rates would fit the graph shown on the right.

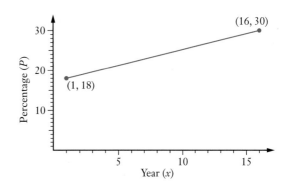

a Determine the equation of this straight line, in the form $P = mx + c$, with x and P as indicated on the graph, and m and c constants.

b Plot a graph of the real percentage figures over the 16 years to show that the points do not form such a straight line and describe any trends the real percentage figures show over time.

8

Transport authorities in a particular country researched vehicle registrations over a number of years to investigate the average age of the vehicles registered for road use. The results were:

Average age of vehicles

Year	1990	1992	1994	1996	1998	2000	2002
Average age (years)	7.6	7.8	8.4	9.1	9.6	10.2	10.5

Year	2004	2006	2008	2010	2012	2014
Average age (years)	10.6	10.7	10.7	10.6	10.4	9.9

a Display this data as a time series graph.

b Is the data suitable for linear regression? Justify your answer.

c Use your graph to predict the average age of vehicles in this country for the years 1999 and 2020, commenting in each case about the likely reliability of your prediction.

9

Confirm that the following paired values are not suited to linear regression.

t	1	2	3	4	5	6	7	8	9	10
Y	5	20	44	96	149	217	310	410	525	661

Explore the ability of graphic calculators to determine non-linear regression models to obtain a model for the above data of the form $Y = a \times t^b$.

THE CHANGING SEASONS

Quite a number of middle-aged or elderly people have been heard to comment that they feel the seasons have been changing over the years. For example, some feel that January tended to be hotter in the past. Are they correct or is it that they are only remembering the more memorable summers?

The table below shows information about January temperatures recorded at a particular location in Perth over a period of approximately 70 years. Does the information given in the table suggest that, during the period of time the statistics cover, summer temperatures in Perth were showing some changing pattern?

Write a brief report justifying your opinion and include suitable graphs based on the information in the table below.

Temperature statistics for the month of January 1946–2014

Year	1946	1948	1950	1952	1954
Mean daily max temp (°C)	28.2	30.9	32.7	31.3	31.0
Mean daily min temp (°C)	13.3	15.3	16.8	14.8	17.6

Year	1956	1958	1960	1962	1964
Mean daily max temp (°C)	33.8	33.1	28.2	35.6	30.6
Mean daily min temp (°C)	17.1	18.2	14.8	19.5	16.7

Year	1966	1968	1970	1972	1974
Mean daily max temp (°C)	31.1	30.5	29.3	32.7	34.2
Mean daily min temp (°C)	17.4	17.5	16.7	16.5	17.4

Year	1976	1978	1980	1982	1984
Mean daily max temp (°C)	30.8	34.0	34.0	29.4	32.4
Mean daily min temp (°C)	16.9	18.4	18.3	15.9	16.8

Year	1986	1988	1990	1992	1994
Mean daily max temp (°C)	33.5	30.2	29.1	33.5	30.9
Mean daily min temp (°C)	18.6	15.5	16.1	19.2	17.1

Year	1996	1998	2000	2002	2004
Mean daily max temp (°C)	31.5	32.6	30.7	31.8	32.4
Mean daily min temp (°C)	16.8	17.3	17.8	16.4	17.6

Year	2006	2008	2010	2012	2014
Mean daily max temp (°C)	29.9	33.8	35.0	33.4	32.9
Mean daily min temp (°C)	16.8	18.8	18.8	19.7	17.8

[Source of data: Bureau of Meteorology.]

RESEARCH AND ANALYSIS

Research and obtain appropriate time series data to answer one of the following questions, or negotiate with your teacher an alternative question of your choice that will similarly require an analysis of time series data.

Your answer should include the data you obtained, suitably presented, an explanation of the analysis of that data and a reasoned conclusion that addresses the question.

- When is the world record for running 100 metres likely to fall to 9 seconds?

- When is the population of the world likely to reach 10 billion?

- What is the population of China likely to be in the year 2050?

- What trends does Australia's unemployment rate show over time?

- How has the share price of a particular company of your choice varied with time and what might this suggest for the future price?

- How have house prices, or rental costs, varied over time in your area and what does this suggest for future price, or rental, movements?

- How has the price of farming commodities varied with time and what does this suggest for future price movements?

- How has the volume of air traffic changed with time and what does this suggest for future air traffic volume?

- How have national or worldwide annual sales of new cars changed with time and what does this suggest for future numbers of sales?

- How have the estimated numbers of some endangered species of animal varied with time and what does this suggest for the future of the species?

Miscellaneous exercise one

This miscellaneous exercise may include questions involving the work of this chapter and the ideas mentioned in the Preliminary work section at the beginning of the book.

1 Increase $2050 by 13%.

2 Decrease $12 600 by 74%.

3 Determine the equation of each of the straight lines A to F shown in the graph on the right.

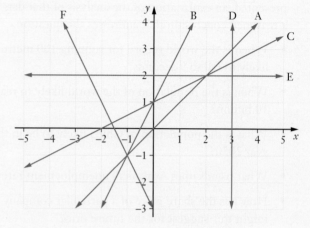

4 Find the first five terms of each of the following recursively defined sequences.

a $T_1 = 25$, $\quad T_{n+1} = T_n + 2$ **b** $T_1 = 32$, $\quad T_{n+1} = T_n - 1.5$

c $T_1 = 64$, $\quad T_{n+1} = 1.5 \times T_n$ **d** $T_1 = 4000$, $\quad T_n = T_{n-1} \div 2$

e $T_1 = 5$, $\quad T_n = 2T_{n-1} + 7$ **f** $T_1 = 16$, $\quad T_{n+1} = 1.5T_n - 4$

5 Population changes

The table on the right shows the population of Perth, and of Western Australia as a whole, in each of the census years from 1911 to 1991.

Make a table of your own showing Perth's population as a percentage of the WA population for each of these census years.

a Plot these percentages as a line graph, with years on the horizontal axis, and comment on the trends your graph shows.

b Use your graph to suggest Perth's population as a percentage of the WA population for **i** 1940 and **ii** 2020, in each case commenting on the likely reliability of your estimates.

Try to find more recent data for the populations of Perth and of Western Australia, say for the census years 2011 and 2016, and comment on how they compare with the data given here.

Population of WA and Perth

Year	WA	Perth
1911	282 114	116 181
1921	332 732	170 213
1933	438 852	230 340
1947	502 480	302 968
1954	639 771	395 049
1961	736 629	475 398
1966	836 673	558 821
1971	1 030 469	703 199
1976	1 178 340	832 760
1981	1 300 600	922 017
1986	1 459 011	1 050 120
1991	1 636 067	1 188 762

[Source of data: Australian Bureau of Statistics]

Moving averages and seasonal effects

- Making underlying trends more apparent
- Moving averages
- General smoothing of time series data
- Use of spreadsheets
- Centred moving averages
- Quantifying the seasonal effect
- Deseasonalising, or seasonally adjusting, the data
- Making predictions
- Miscellaneous exercise two

Situation

During 2014 the owners of a particular tourist attraction decided to use the data regarding the number of visitors the centre has had in recent years to identify trends and to indicate likely visitor figures for the future. The owners have access to the four-monthly visitor totals for 2011, 2012 and 2013 and also for the first four months of 2014. These figures are shown in the table below.

Year	Period of time	Visitors (thousands)
2011	During 1st 4 months	45
	During 2nd 4 months	26
	During 3rd 4 months	34
2012	During 1st 4 months	52
	During 2nd 4 months	37
	During 3rd 4 months	42
2013	During 1st 4 months	64
	During 2nd 4 months	41
	During 3rd 4 months	50
2014	During 1st 4 months	69

- Whilst these visitor figures seem to jump around somewhat, is the underlying trend one of
    increasing numbers of visitors,
    decreasing numbers of visitors,
  or   a reasonably constant number of visitors?

- If the next figure, i.e. for the 2nd period of 4 months of 2014, turns out to be 58, this is clearly a drop from the 69 recorded for the 1st four months of 2014 but is it cause for concern?

The situation above asked you to do two things:

1. Attempt to identify an underlying trend in data that may initially appear rather erratic.

2. Think about the significance of a value beyond those initially given.

In this chapter we will consider

- making underlying trends more apparent,

and - making predictions,

for the sort of fluctuating data given above.

Making underlying trends more apparent

The graph below shows the data from the situation on the previous page.

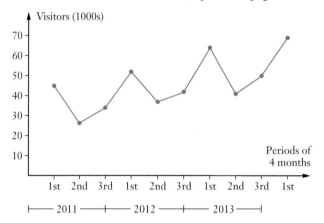

The graph shows the **seasonal nature** of the data with three 'seasons' making up one cycle.

The peaks occur at the 1st, 4th, 7th and 10th points plotted and the troughs at the 2nd, 5th and 8th points plotted. From this pattern we would certainly expect the next value, i.e. for the 2nd 4-month period of 2014, to be a trough. A figure of 58 for this 4-month period would therefore not automatically be a cause for concern unless the drop were more than was expected.

(Note: Seasonal variation has a constant time period. A repeating pattern without a regular frequency of occurrence would be called cyclical rather than seasonal.)

The graph indicates the general upward trend in the data as shown by the orange line below.

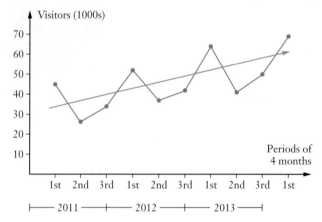

The placement of this line is just a 'guesstimate' based on 'eye-balling' the graph and guessing an approximate line of best fit.

We could obtain a line of best fit using the linear regression techniques that you have encountered previously but, with the original data being quite widely spread, and with an obvious seasonal nature, it is better to first *smooth* the graph by attempting to remove the seasonal variation.

We could smooth the data by taking an average attendance for each of the three years but this will reduce 10 data points down to just 3. Instead, because the given data has a three-point pattern to it we create a **3-point moving average**.

ISBN 9780170395069

Moving averages

A 3-point moving average is found by first averaging the number of visitors for the
1st, 2nd and 3rd data points,
then averaging the   2nd, 3rd and 4th data points,
then averaging the   3rd, 4th and 5th data points, etc.

In this way we move down the data points and each average will be formed using a 1st 4 months figure, a 2nd 4 months figure, and a 3rd 4 months figure.

The first three 3-point moving averages are shown below (rounded to 1 decimal place when appropriate). Note carefully the placement of the moving average (MA) figures.

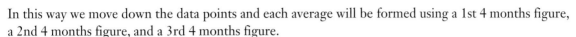

Yr	Months			MA
11	1st 4	45		
	2nd 4	26	→	35
	3rd 4	34		
12	1st 4	52		
	2nd 4	37		
	3rd 4	42		

Yr	Months			MA
11	1st 4	45		
	2nd 4	26		
	3rd 4	34	→	37.3
12	1st 4	52		
	2nd 4	37		
	3rd 4	42		

Yr	Months			MA
11	1st 4	45		
	2nd 4	26		
	3rd 4	34		
12	1st 4	52	→	41
	2nd 4	37		
	3rd 4	42		

The complete table is shown on the right.

Adding these values to the graph (see the orange dashed line below) we see that this moving average technique has smoothed the data considerably.

Year	Months	Visitors	3-pt MA
2011	1st 4	45	–
	2nd 4	26	35
	3rd 4	34	37.3
2012	1st 4	52	41
	2nd 4	37	43.7
	3rd 4	42	47.7
2013	1st 4	64	49
	2nd 4	41	51.7
	3rd 4	50	53.3
2014	1st 4	69	–

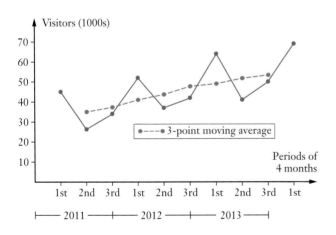

Note

The moving average has smoothed the seasonal component in the data, leaving the trend component more apparent.

The *three*-point moving average achieved this smoothing because the original data had a *three*-point pattern to it. Had the original data had a *four*-point pattern, as we might expect if we were collecting data every quarter year, we would use a *four*-point moving average as shown in example 1 on the next page. If data was collected each month we might expect a 12-point pattern. If data was collected every day we might expect a 7-point pattern.

ISBN 9780170395069

EXAMPLE 1

To monitor the planned eradication of a particular disease in cattle all vets were required to notify the government of any cattle diagnosed as having the disease. The data collected was totalled every three months (or quarter of a year) and is shown on the right.

a Calculate the 4-point moving averages.

b On a single graph plot both the original data and the 4-point moving averages.

c Is the underlying trend decreasing, steady or increasing?

Year	Period of time	Number of cattle
2011	1st quarter	60
	2nd quarter	67
	3rd quarter	47
	4th quarter	38
2012	1st quarter	52
	2nd quarter	53
	3rd quarter	43
	4th quarter	28
2013	1st quarter	42
	2nd quarter	46
	3rd quarter	34
	4th quarter	23
2014	1st quarter	28
	2nd quarter	32

Solution

a The 4-point moving averages are shown in the table below. Again note carefully the placement of these figures.

Year	Time	No. of cattle	4-pt MA
2011	1st quarter	60	
	2nd quarter	67	
			→ 53
	3rd quarter	47	
			51
	4th quarter	38	
			47.5
2012	1st quarter	52	
			46.5
	2nd quarter	53	
			44
	3rd quarter	43	
			41.5
	4th quarter	28	
			39.75
2013	1st quarter	42	
			37.5
	2nd quarter	46	
			36.25
	3rd quarter	34	
			32.75
	4th quarter	23	
			29.25
2014	1st quarter	28	
	2nd quarter	32	

b The initial data and the 4-point moving averages are shown graphed below.

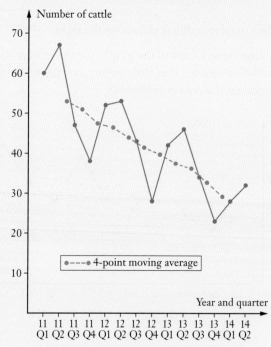

c The number of cattle diagnosed with the disease is showing a decreasing trend.

Exercise 2A

For each of numbers **1** to **6** suggest the most appropriate moving average to use to smooth the given set of data. (i.e. 3-point, 4-point, ? point.)

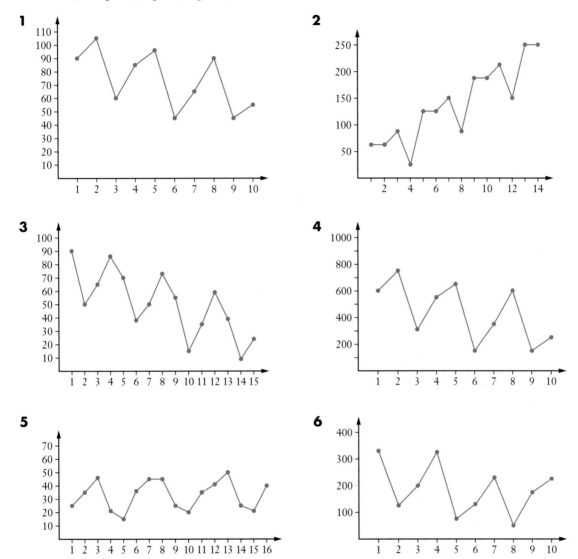

Note for the remaining questions of this exercise:

Some calculators and computer programs, when given the original data, can be programmed to list moving averages directly. However, even if your have access to such capability, do the remaining questions of this exercise without using this facility, to ensure that you fully understand the process.

7 A drama group put on a particular play three times a week for five weeks.

The capacity of the theatre they use for the production is 500 and the numbers attending over the five-week 'season' are given below right.

a Plot this data as a graph with the performance number on the horizontal axis and attendance on the vertical axis.

b Add to your graph a plot of the data after it has been smoothed using a three-point moving average.

c Do the attendance figures indicate that a sixth week of performances would have been a good idea?

Performance number	Week	Day	Attendance
1		Thur	210
2	1	Fri	320
3		Sat	370
4		Thur	270
5	2	Fri	440
6		Sat	450
7		Thur	420
8	3	Fri	500
9		Sat	500
10		Thur	450
11	4	Fri	500
12		Sat	500
13		Thur	350
14	5	Fri	410
15		Sat	430

8 To monitor the planned eradication of a particular disease in pigs all vets were required to notify the government of any pig diagnosed as having the disease. The data collected was totalled every three months (or quarter of a year) and is shown on the right.

a Calculate the 4-point moving averages.

b On a single graph plot both the original data and the 4-point moving averages.

c Is the underlying trend decreasing, steady or increasing?

Year	Period of time	Number of pigs
1	1st quarter	29
	2nd quarter	31
	3rd quarter	21
	4th quarter	25
2	1st quarter	28
	2nd quarter	24
	3rd quarter	21
	4th quarter	23
3	1st quarter	26
	2nd quarter	24
	3rd quarter	19
	4th quarter	20
4	1st quarter	25
	2nd quarter	19

ISBN 9780170395069

9 The table shown relates to the power bills received by a household every two months for a period of three years.

a Determine values for the spaces marked A, B, C, D and E.

b On a single graph plot both the bill amounts for the three years and the six-point moving averages.

c Is the underlying trend decreasing, steady or increasing?

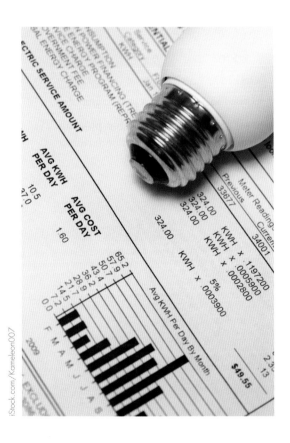

	Year	2 month period	Bill ($)	6-pt MA
1	2012	1st 2 months	238	
				–
2		2nd 2 months	251	
				–
3		3rd 2 months	283	
				258
4		4th 2 months	306	
				261
5		5th 2 months	274	
				A
6		6th 2 months	196	
				B
7	2013	1st 2 months	256	
				C
8		2nd 2 months	263	
				D
9		3rd 2 months	325	
				284
10		4th 2 months	342	
				287
11		5th 2 months	280	
				288
12		6th 2 months	238	
				287
13	2014	1st 2 months	274	
				289
14		2nd 2 months	269	
				290
15		3rd 2 months	319	
				293
16		4th 2 months	354	
				–
17		5th 2 months	286	
				–
18		6th 2 months	E	

General smoothing of time series data

Even if there is no obvious seasonal nature to time series data we can still use a moving average to smooth out any **irregular fluctuations** and **unsystematic peaks and troughs**. For example consider the following table showing the total wool production in Australia, in tonnes, every five years from 1915 to 2010.

Year	1915	1920	1925	1930	1935
Total production (tonnes)	291 300	300 800	330 700	425 600	460 500

Year	1940	1945	1950	1955	1960
Total production (tonnes)	511 400	460 000	516 800	580 500	760 500

Year	1965	1970	1975	1980	1985
Total production (tonnes)	807 900	912 100	729 200	658 200	752 200

Year	1990	1995	2000	2005	2010
Total production (tonnes)	1 049 800	679 400	641 500	475 200	352 700

(Based on Australian Bureau of Statistics data.)

Plotting this data below shows no particular systematic, calendar related movements but plotting the moving average data (shown dashed) still has a smoothing effect and can make an underlying trend clearer.

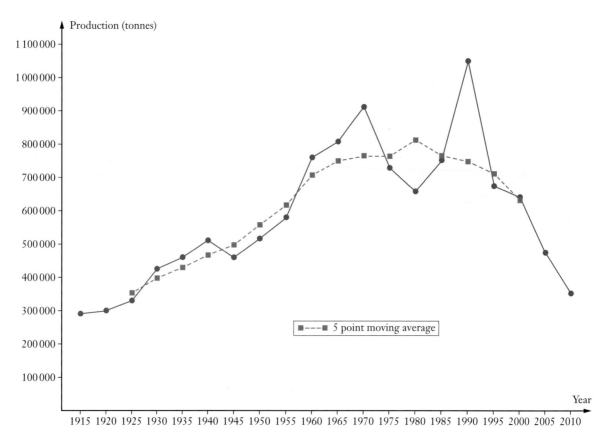

ISBN 9780170395069

Use of spreadsheets

Spreadsheets can be particularly suitable for calculating moving averages and displaying the results graphically. The information from the previous situation involving wool production in Australia can be put into a spreadsheet, the moving averages can be calculated and the data displayed graphically, as shown below.

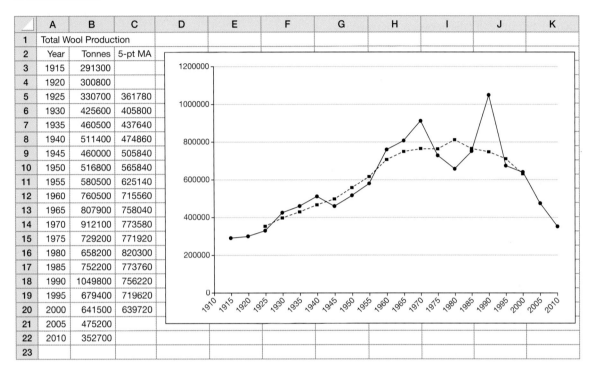

	A	B	C	D	E	F	G	H	I	J	K
1	Total Wool Production										
2	Year	Tonnes	5-pt MA								
3	1915	291300									
4	1920	300800									
5	1925	330700	361780								
6	1930	425600	405800								
7	1935	460500	437640								
8	1940	511400	474860								
9	1945	460000	505840								
10	1950	516800	565840								
11	1955	580500	625140								
12	1960	760500	715560								
13	1965	807900	758040								
14	1970	912100	773580								
15	1975	729200	771920								
16	1980	658200	820300								
17	1985	752200	773760								
18	1990	1049800	756220								
19	1995	679400	719620								
20	2000	641500	639720								
21	2005	475200									
22	2010	352700									
23											

Try to create your own spreadsheets to answer some of the questions in this chapter.

Note: You may find that a spreadsheet offers the facility of adding a trend line to a graph, and one of the trend line options may be a moving average line for which you can set the period, i.e. a 3-point, 4-point, … N-point moving average. However it may be the case that the moving average points are plotted at the end of the period not at the centre, as we tend to show them. Plotting values at the end of the values that are used to form the moving averages produces the same line as shown dashed above, but moved to the right. In such a line the same values are displayed 'later'. The line is said to be showing the *trailing* moving averages. In this text we will not use this line and will instead always plot our moving averages at the centre of the time values used to form it, as shown above.

However, that is not to say that attaching a moving average to the end of the period used to form it cannot be useful in some circumstances. Suppose for example that you were an athlete monitoring your times in training. Today you might well form a three-point average involving today's time and the times of the previous two days. In your training diary you could well record today's figure as being the average of today's time and that of the previous two days. I.e. you would be recording the three-point moving average at the end of each three-day period used to calculate it. But, as stated above, when plotting moving averages in this text in order to smooth data we will plot at the centre of the time interval used to calculate it.

Centred moving averages

Consider again the calculation of four-point moving averages encountered previously involving data collected with regard to a disease in cattle, as shown on the right. Note the moving average figure does not align with a particular quarter but instead falls between quarters.

This 'misalignment' of original data and moving average figures will occur in an N-point moving average when N is even. For some uses of moving averages it is more convenient if the two quantities do align. Hence it is quite common to give an N-point moving average *centred* when N is even. This centring simply involves taking the average of two values, as the following table shows for the cattle disease data.

Year	Quarter	No. of cattle	4-pt MA
2011	1st	60	
	2nd	67	
			→ 53
	3rd	47	
	4th	38	

t	Year	Quarter	No. of cattle	4-pt MA	Centred 4-pt MA
1	2011	1st	60		
2		2nd	67		
			→	53	
3		3rd	47		→ 52
				51	
4		4th	38		49.25
				47.5	
5	2012	1st	52		47
				46.5	
6		2nd	53		45.25
				44	
7		3rd	43		42.75
				41.5	
8		4th	28		40.625
				39.75	
9	2013	1st	42		38.625
				37.5	
10		2nd	46		36.875
				36.25	
11		3rd	34		34.5
				32.75	
12		4th	23		31
				29.25	
13	2014	1st	28		
14		2nd	32		

ISBN 9780170395069

Note: • We can go directly from the initial data to the centred four-point moving average if we take the **five** values that contribute to the centred average and sum 'half the first, half the last, and the middle three' and then divide by four. Applying this idea to the previous example, the first five values give

$$\frac{30 \text{ (i.e. half of 60)} + 67 + 47 + 38 + 26 \text{ (i.e. half of 52)}}{4} = 52, \text{ as required.}$$

The next set of five values give:

$$\frac{33.5 + 47 + 38 + 52 + 26.5}{4} = 49.25, \text{ as required.}$$

• Use a, b, c, d and e to prove that the above method gives the same result as is obtained by first obtaining the uncentred 4-point averages.

• What would be the rule for centring six-point averages in this way?

Exercise 2B

1 The table below shows the number of cans of soft drink a lunch bar sells in each three months of a three-year period.

a Why would determining 4-point moving averages be suitable for this data?

b Determine the values A, B, C, D, E, F and G.

t	Year	Quarter	Cans sold	4-pt MA	Centred 4-pt MA
1	One	1st	8312		–
				–	
2		2nd	5783		–
				7006	
3		3rd	5727		6983
				6960	
4		4th	8202		A
				6920	
5	Two	1st	8128		6906
				B	
6		2nd	5623		6840
				C	
7		3rd	5615		D
				6738	
8		4th	7786		6724
				6710	
9	Three	1st	7928		6698
				6686	
10		2nd	E		G
				F	
11		3rd	5519		–
				–	
12		4th	7594		–

2 The table shows the number of client interviews an accountant conducts in each 2-month period over three years.

  a Determine the values A, B, C, D E, F, G, H, I and J.

t	Year	2-month period	Units	6-pt MA	Centred 6-pt MA
1	1	1st	142		
2		2nd	235		
3		3rd	295		
				175	
4		4th	141		A
				178	
5		5th	80		C
				B	
6		6th	157		D
				184	
7	2	1st	160		183.5
				E	
8		2nd	247		F
				185	
9		3rd	319		185.5
				186	
10		4th	135		185
				184	
11		5th	92		181.5
				179	
12		6th	163		181
				183	
13	3	1st	148		185.5
				188	
14		2nd	217		I
				H	
15		3rd	343		J
				193	
16		4th	165		
17		5th	104		
18		6th	G		

  b Reproduce the centred 6-point moving average figures using a spreadsheet showing only two columns, the raw data 'units' column (with your value for G included) and the centred moving average column. (See the note at the top of the previous page.)

ISBN 9780170395069

3 Determine the moving averages for the information given below, which shows the tonnage of a particular crop produced in each season, over two years. Take MAs to mean Moving Averages and CMAs to mean Centred Moving Averages.

	Year 1				Year 2		
Summer	Autumn	Winter	Spring	Summer	Autumn	Winter	Spring
1228	364	640	1220	1436	276	752	1132

4-point MAs

4-point CMAs

4 The monthly average price per kilogram of a particular type of fruit was monitored over a period of four years. The data obtained is shown below.

Year 1					
January	February	March	April	May	June
$2.90	$2.70	$1.90	$1.80	$1.75	$1.90
July	August	September	October	November	December
$1.90	$1.95	$2.00	$2.15	$2.60	$3.00

Year 2					
January	February	March	April	May	June
$3.30	$2.40	$1.90	$1.95	$2.00	$1.95
July	August	September	October	November	December
$2.00	$2.10	$2.25	$2.35	$2.60	$2.80

Year 3					
January	February	March	April	May	June
$3.20	$3.60	$2.60	$2.20	$2.05	$2.00
July	August	September	October	November	December
$2.00	$2.05	$2.10	$2.20	$2.60	$3.10

Year 4					
January	February	March	April	May	June
$3.20	$2.65	$2.05	$1.95	$2.05	$2.20
July	August	September	October	November	December
$2.40	$2.50	$2.60	$2.90	$3.40	$4.30

a Why does it make sense to consider the 12-point moving averages for this data?

b Either manually, or with the assistance of a graphic calculator, spreadsheet or computer program display the original data, and the 12-point centred moving averages, on a single graph.

c Is the underlying trend decreasing, steady or increasing?

5 If the population of a region increases we might expect the number of criminal incidents reported by the police for that region to similarly increase.

For crimes reported by New South Wales police and classified as '*Robbery with a firearm*' the numbers reported annually from 1990 to 2013 were as follows (rounded to the nearest multiple of 5):

Number of incidents classified as 'Robbery with a firearm' recorded by NSW police, 1990 to 2013

Year	Number	Year	Number	Year	Number
1990	980	1998	1120	2006	605
1991	1265	1999	890	2007	595
1992	1130	2000	805	2008	420
1993	875	2001	1105	2009	520
1994	520	2002	890	2010	445
1995	785	2003	835	2011	400
1996	1015	2004	710	2012	380
1997	1400	2005	520	2013	325

(© State of New South Wales through the Department of Justice and reproduced with the approval of the NSW Bureau of Crime Statistics and Research.)

a Which would smooth the data more, a three-point moving average or a five-point moving average?

b Display on a single graph the original figures, and whichever of the three- or five-point moving averages you felt would smooth the data more.

c Comment on any trends shown.

Moving averages and finance

Moving averages are sometimes used by people monitoring the movement of share prices. A 30-day moving average of share prices will be less influenced by a sudden change in the share price than say a 5-day moving average would be. By considering and comparing different moving averages a stock market analyst can attempt to identify trend and predict future movement in share prices.

In some cases the analyst might 'weight' the moving average so that the more recent share prices have more influence on the average than a price from a few weeks ago. Whilst this idea of a weighted moving average is beyond the scope of this course, do a bit of research on the internet to see what some financial sites say about the use of moving averages.

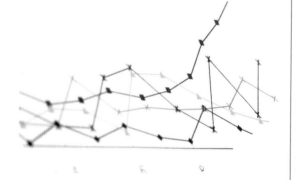

iStock.com/alubalish

ISBN 9780170395069

Quantifying the seasonal effect

Let us consider again the situation from the beginning of this chapter concerning visitor numbers to a particular tourist attraction. These figures are shown again below.

Year	Period of time	Visitors (thousands)
2011	During 1st 4 months	45
	During 2nd 4 months	26
	During 3rd 4 months	34
2012	During 1st 4 months	52
	During 2nd 4 months	37
	During 3rd 4 months	42
2013	During 1st 4 months	64
	During 2nd 4 months	41
	During 3rd 4 months	50
2014	During 1st 4 months	69

Notice that there seems to be a '1st 4 months effect' that leads to a higher number of visitors attending the tourist attraction in the first four months than in the other two periods of four months in each year.

Consider the figures for 2011:

$$\text{Total number of visitors in 2011} = 45\,000 + 26\,000 + 34\,000$$
$$= 105\,000$$

$$\text{Average number of visitors per 4-month period} = 105\,000 \div 3$$
$$= 35\,000$$

The 1st 4 month number, 45 000, as a percentage of this average figure, is 128.6% (1 decimal place).

The 2nd 4 month number, 26 000, as a percentage of this average figure, is 74.3% (1 decimal place).

The 3rd 4 month number, 34 000, as a percentage of this average figure, is 97.1% (1 decimal place).

Hence we could say that for 2011 the 1st 4 months have a seasonal effect that lifts the attendance by 29% of the average, the 2nd 4 months have a seasonal effect that drops the attendance by 26% of the average (74% is 26% below 100%) and the 3rd 4 months have a seasonal effect that drops the attendance by 3% of the average.

However, with the figures for 2012 and 2013 available it would be better to carry out the above process for all three years and then average the results. This is indeed how we will determine a measure of the effect each season has – called the **seasonal effect**, **seasonal component** or **seasonal index**.

The three **seasonal indices** for the above data are calculated on the next page.

Year	Period of time	Visitors (1000s)	4-month mean for the year (1000s)	Visitor numbers as percentage of 4 month mean
2011	During 1st 4 months	45	35	128.57%
	During 2nd 4 months	26		74.29%
	During 3rd 4 months	34		97.14%
2012	During 1st 4 months	52	43.667	119.08%
	During 2nd 4 months	37		84.73%
	During 3rd 4 months	42		96.18%
2013	During 1st 4 months	64	51.667	123.87%
	During 2nd 4 months	41		79.35%
	During 3rd 4 months	50		96.77%
2014	During 1st 4 months	69		

Calculation of seasonal effect, seasonal component or seasonal index:

	1st 4 months	2nd 4 months	3rd 4 months
2011	128.57%	74.29%	97.14%
2012	119.08%	84.73%	96.18%
2013	123.87%	79.35%	96.77%
Mean (nearest %)	124%	79%	97%

The seasonal index for the first 4 months of a year is 124% (or 1.24), the seasonal index for the second 4 months of a year is 79% (or 0.79) and the seasonal index for the third 4 months of a year is 97% (or 0.97). This means that visitor numbers for the first four months of a year tend to be 24% above the average 4 month numbers, for the second four months they tend to be 21% below the average 4 month numbers and for the third 4 months of a year they tend to be 3% below the average 4 month numbers.

Note:
- This particular method of determining the **seasonal indices** is called the **average percentage method**. There are a number of other methods that are used to quantify seasonal effects in time series data but in this unit we will use this average percentage method.

- The average of the seasonal indices should be 100%, i.e. with three seasons as shown above the total should be 300%. This is the case for the above indices, $124 + 79 + 97 = 300$. If small errors due to rounding make an average not equal to 100, suitable adjustment could be made.

- On occasions, though rarely in this text, the average percentage method uses the median rather than the mean of the percentages. From the above table this would give indices of 123.87%, 79.35% and 96.77% (i.e. in this case it would make little difference, but that would not always be the case). In this text, when asked for the seasonal indices to be calculated using the average percentage method, the reader should assume the mean is to be used.

ISBN 9780170395069

Read through the following which shows the calculation of seasonal indices for data involving six seasons and make sure you understand the calculations involved.

t	Year	2-month period	Units	2-month mean for the year	Percentage of 2-month mean
1	1	1st	284	352	80.68%
2		2nd	463		131.53%
3		3rd	595		169.03%
4		4th	279		79.26%
5		5th	168		47.73%
6		6th	323		91.76%
7	2	1st	318	370	85.95%
8		2nd	493		133.24%
9		3rd	635		171.62%
10		4th	268		72.43%
11		5th	188		50.81%
12		6th	318		85.95%
13	3	1st	302	381	79.27%
14		2nd	428		112.34%
15		3rd	678		177.95%
16		4th	308		80.84%
17		5th	211		55.38%
18		6th	359		94.23%

	1st 2 months	2nd 2 months	3rd 2 months	4th 2 months	5th 2 months	6th 2 months
1st year	80.68%	131.53%	169.03%	79.26%	47.73%	91.76%
2nd year	85.95%	133.24%	171.62%	72.43%	50.81%	85.95%
3rd year	79.27%	112.34%	177.95%	80.84%	55.38%	94.23%
Mean	82.0%	125.7%	172.9%	77.5%	51.3%	90.6%

The seasonal indices for the 1st 2 months, 2nd 2 months etc. are

1st 2 months: 82.0% (or 0.820),	2nd 2 months: 125.7% (or 1.257),
3rd 2 months: 172.9% (or 1.729),	4th 2 months: 77.5% (or 0.775),
5th 2 months: 51.3% (or 0.513),	6th 2 months: 90.6% (or 0.906).

Notice that $82.0 + 125.7 + 172.9 + 77.5 + 51.3 + 90.6 = 600$, as we would want for data involving six seasons as then the average index is 100. However, were we to round each index to an integer we would have indices of 82, 126, 173, 78, 51 and 91 which give a total of 601, the 'extra 1' being introduced due to the rounding carried out. To avoid this situation we could choose to round the 77.5 down to 77, to give indices of 82, 126, 173, 77, 51 and 91. However in real applications we would probably be going on to use these seasonal indices for further calculations, so we would be able to use more accurate values stored in the cell of a spreadsheet, and would leave all rounding until the final calculation is completed.

Exercise 2C

1 Three of the seasonal indices for data with a quarterly seasonal pattern are as follows:

Seasonal index for the first quarter:	113%
Seasonal index for the second quarter:	105%
Seasonal index for the third quarter:	84%

Determine the seasonal index for the fourth quarter.

2 Six of the seasonal indices for data with a daily seasonal pattern are as follows:

Seasonal index for Monday:	73%
Seasonal index for Tuesday:	78%
Seasonal index for Thursday:	89%
Seasonal index for Friday:	102%
Seasonal index for Saturday:	143%
Seasonal index for Sunday:	121%

Determine the seasonal index for Wednesday.

3 A company expects the annual sales of one of their products next year to be 156 000 and, from past years' experience, anticipates no underlying increasing or decreasing trend, and no irregular fluctuations in sales through the year.

a If there is no monthly seasonal effect on sales of this product roughly how many of the product should the company expect to sell each month?

b In fact the sales are seasonal in nature with the seasonal indices for each month being as follows:

January	88%	February	82%	March	76%
April	74%	May	83%	June	98%
July	104%	August	105%	September	108%
October	111%	November	126%	December	145%

Predict the sales numbers for this product for each month of next year.

4 A company finds that in the first quarter of the year it sells 17 360 units of a particular product.

From the quarterly sales of this product in previous years the company has calculated the seasonal indices for sales of this product per quarter as being as follows:

Seasonal index for the first quarter:	112%
Seasonal index for the second quarter:	104%
Seasonal index for the third quarter:	98%
Seasonal index for the fourth quarter:	86%

Assuming that for this year the only anticipated change in sales from one month to the next is due to this seasonal effect, what do these figures suggest for the likely number of units of this product sold:

a in the whole year?

b in each of the remaining three quarters of the year?

ISBN 9780170395069

Copy and complete each of the following sets of seasonal data, filling in the spaces marked '?'. (Whilst the average of the indices should be 100% do not worry if, due to rounding, this is not quite the case.)

5

Week	Period of time	Attendance	Daily mean for the week	Attendance as % of week's daily mean
1	Friday	2346	2766	84.82%
	Saturday	3143		113.63%
	Sunday	2809		101.55%
2	Friday	2572	?	?
	Saturday	3258		?
	Sunday	2957		100.96%
3	Friday	2987	?	?
	Saturday	3500		?
	Sunday	3206		99.23%

	Friday	Saturday	Sunday
1st week	84.82%	113.63%	101.55%
2nd week	?	?	100.96%
3rd week	?	?	99.23%
Index (1 decimal place)	?	?	100.6%

6

Year	Period of time	Units sold	Quarterly mean for the year	Units as % of year's quarterly mean
1	1st quarter	54 000	52 650	102.56%
	2nd quarter	63 200		120.04%
	3rd quarter	45 200		85.85%
	4th quarter	48 200		?
2	1st quarter	48 500	?	102.65%
	2nd quarter	56 700		?
	3rd quarter	41 300		?
	4th quarter	42 500		?
3	1st quarter	51 800	?	104.28%
	2nd quarter	62 600		?
	3rd quarter	40 500		?
	4th quarter	43 800		?

	1st quarter	2nd quarter	3rd quarter	4th quarter
1st year	102.56%	120.04%	85.85%	?
2nd year	102.65%	?	?	?
3rd year	104.28%	?	?	?
Seasonal index (1 decimal place)	103.2%	?	?	?

7 A local council commences a campaign to encourage people who need to contact the council for various matters to do so via the council's interactive website, thus avoiding the delays experienced by people making phone or direct contact with the council's help desk.

To monitor the success of this venture in reducing the number of calls coming into the help desk, and to be able to predict future staffing needs, the council monitors the number of non-internet phone calls made to the help desk on each day of a three-week period. The results are shown in the table below.

The council wants to know appropriate moving average figures for the three-week period and the seasonal index for each day.

With the assistance of a spreadsheet, construct completed versions of the following tables.

n	Week	Day	No. of calls	5-pt MA	Calls as % of daily average for week
1		Mon	258	–	107.59%
2		Tue	231	–	96.33%
3	One	Wed	215	239.8	
4		Thur	248	235.6	
5		Fri	247	231.0	
6		Mon	237		
7		Tue	208		
8	Two	Wed	184		
9		Thur	232		
10		Fri	223		
11		Mon	232		
12		Tue	194		
13	Three	Wed	183		
14		Thur	209	–	
15		Fri	208	–	

	Week 1	Week 2	Week 3	Seasonal index
Monday	107.59%			
Tuesday	96.33%			
Wednesday				
Thursday				
Friday				

ISBN 9780170395069

Deseasonalising, or seasonally adjusting, the data

Now that we are able to quantify the effect a season has, we can remove this seasonal effect and present the data *deseasonalised* or *seasonally adjusted*. Considering again the visitor numbers for the tourist attraction situation encountered earlier:

Year	Period of time	Visitors (1000s)
2011	During 1st 4 months	45
	During 2nd 4 months	26
	During 3rd 4 months	34
2012	During 1st 4 months	52
	During 2nd 4 months	37
	During 3rd 4 months	42
2013	During 1st 4 months	64
	During 2nd 4 months	41
	During 3rd 4 months	50
2014	During 1st 4 months	69

We calculated earlier that the seasonal index for the 1st 4 months is 124%, or 1.24. Thus the figure of 45 000 visitors includes a 24% 'seasonal increase'. To remove this 24% increase we divide the 45 000 by 1.24 to give 36 thousand, to the nearest thousand.

The seasonal index for the 2nd 4 months is 79%, or 0.79. Thus the figure of 26 000 visitors includes a 21% 'seasonal decrease'. To remove this 21% decrease we divide the 26 000 by 0.79 to give 33 000, to the nearest thousand.

Continuing in this way gives the seasonally adjusted figures shown in the table below:

Year	Period of time	Visitors (1000s)	Seasonally adjusted visitor numbers (1000s)
2011	During 1st 4 months	45	36
	During 2nd 4 months	26	33
	During 3rd 4 months	34	35
2012	During 1st 4 months	52	42
	During 2nd 4 months	37	47
	During 3rd 4 months	42	43
2013	During 1st 4 months	64	52
	During 2nd 4 months	41	52
	During 3rd 4 months	50	52
2014	During 1st 4 months	69	56

Making predictions

Let us suppose that a bank decides to operate a *7-days-a-week phone helpline* to give advice regarding new home loans and that the number of calls this helpline receives in the first three weeks of operation are as shown in the table below.

Can we use these figures to predict the number of calls this helpline is likely to receive on the Monday and Saturday of week 4 (assuming any trends of the first three weeks continue)?

If we plot the data, as shown below, it might appear that any prediction would be difficult given the fluctuating nature of the points.

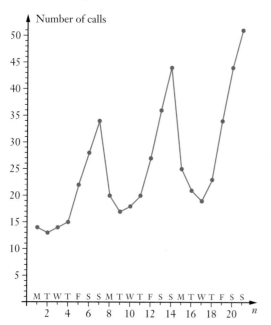

n	Week	Day	No. of calls
1		Mon	14
2		Tue	13
3		Wed	14
4	1	Thur	15
5		Fri	22
6		Sat	28
7		Sun	34
8		Mon	20
9		Tue	17
10		Wed	18
11	2	Thur	20
12		Fri	27
13		Sat	36
14		Sun	44
15		Mon	25
16		Tue	21
17		Wed	19
18	3	Thur	23
19		Fri	34
20		Sat	44
21		Sun	51

However, if we add the 7-point moving average data to this graph, or if we plot the deseasonalised data, both shown on the next page, we have straighter lines, showing less fluctuation. We could then, with more confidence, use linear regression techniques to predict future values (although we know that such prediction would involve extrapolation and so any predicted values would need to be viewed with caution).

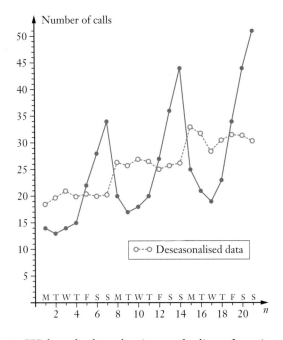

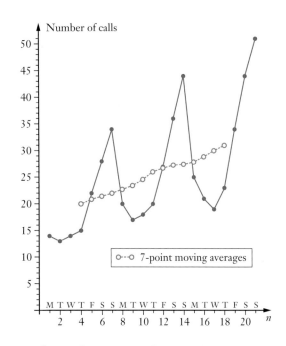

- With each plotted point on the line of moving averages being the average of seven points it is perhaps to be expected that the moving average line shows less variation than the deseasonalised data line. One 'unusual value' will have limited influence in changing the direction of a line when it is but one of seven scores given equal weighting when determining the moving average. Moving averages tend to remove seasonality and individuality. They smooth all variation, not just seasonal variation.

- The deseasonalised data was formed using the following seasonal indices calculated using the average percentage method explained earlier in the chapter.

Monday: 0.76   Tuesday: 0.66   Wednesday: 0.67   Thursday: 0.75

Friday: 1.08   Saturday 1.40   Sunday: 1.68

Now if we are to use either the moving average values, or the deseasonalised values, to determine a line of best fit using linear regression techniques, any prediction we make using such a line of best fit will be a predicted moving average value, or a predicted deseasonalised value. For these predictions to be 'real-life values', rather than values that have been smoothed in some way, we will need to factor back in the seasonal effect.

For example, if we use linear regression to determine a line of best fit for predicting D, the deseasonalised data, given the value of n, the 'day number', where $n = 1$ corresponds to Monday of week 1, we obtain the equation $D = 0.70n + 18.0$.

Thus: When $n = 22$ (i.e. Monday of week 4),   predicted $D = 33.4$.
            When $n = 27$ (i.e. Saturday of week 4),   predicted $D = 36.9$.

Hence to 'factor back in the seasonal effect' the predictions will be as follows:

For Monday of week 4:   Predicted number of calls   = 33.4×0.76,   i.e. approx. 25.
For Saturday of week 4:   Predicted number of calls   = 36.9×1.40,   i.e. approx. 52.

Note: Remember that not all data can be suitably modelled using linear regression. Just because smoothing techniques such as deseasonalising the data or finding the moving averages have been carried out we cannot assume that the resulting data is necessarily showing a linear pattern.

Recall for example the wool production data encountered earlier in this chapter, for which the moving average plot is shown below.

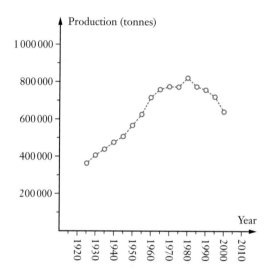

As the graph indicates, applying linear regression techniques to the whole data set would **not** be appropriate.

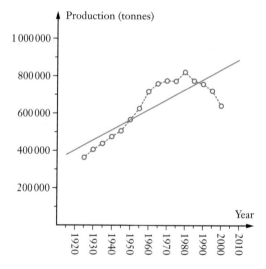

The straight line given by using linear regression is shown above and is not a very good fit for the data.

Remember:

$$\text{Deseasonalised data (or seasonally adjusted data)} = \frac{\text{Real data}}{\text{Seasonal index}}$$

and so

$$\text{Real data} = \text{Deseasonalised data} \times \text{Seasonal index}$$

Note: In this unit we are using a *multiplicative model* for determining the seasonal indices. Hence the statements at the base of the previous page are true. In some cases, though not in this unit, an additive model could be used and in such cases seasonal adjustment occurs by addition or subtraction, not by multiplication.

Exercise 2D

1 A company achieves profits of $25 400 in June of one year. If the June profit figures for this company have a seasonal index of 114% what would be the deseasonalised, or seasonally adjusted, profit for June? (Give your answer to the nearest $100.)

2 From an analysis of sales from earlier years, a company statistician determines that the number of units of a particular product the company sells follows a seasonal pattern with the following quarterly indices:

First quarter	0.65	Second quarter	0.76
Third quarter	0.82	Fourth quarter	1.77

Deseasonalise a figure of 132 000 units sold in the second quarter, giving your answer to the nearest 100 units.

3 a It is announced that a particular region had 65 470 people unemployed in January 'seasonally adjusted'. If the seasonal index for January is 106.2% what was the 'real number' unemployed for this region in January? (Give your answer to the nearest ten people.)

b In May of the same year the region had 67 140 people unemployed. If the seasonal index for May is 96% what would the real figure of 67 140 be reported as after 'seasonal adjustment'? (Again give your answer to the nearest ten people.)

4 Projecting forward using smoothed data a company analyst finds that the predicted sales figure for January will be 12 500 units. However this has not allowed for the January seasonal effect. If the seasonal index for January is 1.18 what does this predict for the number of units sold in January?

5 Projecting forward using smoothed data the statistically astute owner of a number of orchards predicts that the orchards will produce 16 000 kg of a particular fruit this summer, but he knows that this prediction has not yet allowed for the seasonal effect of summer. If the summer seasonal index is 1.48, what does this suggest the actual weight of this fruit from the orchards will be this summer?

6 Let us suppose that the four-monthly totals for short-stay visitors arriving in a particular country, collected over a four-year period, are as shown in the table below.

    a Calculate the seasonal indices for each of the seasons Jan → Apr, May → Aug, Sept → Dec. (Give indices as percentages correct to 2 decimal places, e.g. 105.72%.)

    b Deseasonalise the numbers in the arrivals column.

n	Year	Months	Arrivals (millions)
1		Jan → Apr	1.83
2	1	May → Aug	1.29
3		Sept → Dec	1.98
4		Jan → Apr	1.87
5	2	May → Aug	1.41
6		Sept → Dec	2.06
7		Jan → Apr	1.93
8	3	May → Aug	1.28
9		Sept → Dec	2.16
10		Jan → Apr	2.18
11	4	May → Aug	1.52
12		Sept → Dec	2.33

7 A show is held each year over a long weekend, starting on the Friday and finishing on the public holiday Monday. The attendance figures for four consecutive years are shown in the table below.

    a Calculate the seasonal index for each of the four days. (Give your answers as a decimal and correct to 4 decimal places, e.g. 1.1234.)

    b Seasonally adjust the numbers in the attendance column. (Answer to the nearest 10.)

Year	Day	Attendance
	Friday	9200
1	Saturday	14840
	Sunday	16260
	Monday	16850
	Friday	9810
2	Saturday	15790
	Sunday	17270
	Monday	17800
	Friday	9450
3	Saturday	15220
	Sunday	17200
	Monday	16810
	Friday	9840
4	Saturday	16900
	Sunday	18040
	Monday	16750

8 The table below shows the evening takings for a new take-away fish and chip shop, each evening for the first three weeks of operation.

t	Week	Day	Takings	Deseasonalised weekly takings (D)
1		Monday	$537	
2		Tuesday	$618	
3		Wednesday	$372	
4	1	Thursday	$770	
5		Friday	$937	
6		Saturday	$915	
7		Sunday	$492	
8		Monday	$605	
9		Tuesday	$730	
10		Wednesday	$410	
11	2	Thursday	$880	
12		Friday	$1085	
13		Saturday	$1003	
14		Sunday	$495	
15		Monday	$705	
16		Tuesday	$804	
17		Wednesday	$420	
18	3	Thursday	$840	
19		Friday	$1240	
20		Saturday	$1270	
21		Sunday	$531	

a Given that the seasonal indices are as shown below, calculate the entries for the deseasonalised weekly takings column, to the nearest dollar.

Monday	Tuesday	Wednesday	Thursday	Friday	Saturday	Sunday
82.42%	96.07%	53.94%	111.87%	145.52%	141.94%	68.24%

b Use a graphic calculator or computer spreadsheet to view a graph showing the deseasonalised figures, D, on the vertical axis plotted against t on the horizontal axis (t is as shown in the above table) to confirm that for the variables t and D linear regression would be appropriate.

c Use t and the deseasonalised data, D, to determine the equation of the regression line $D = A + Bt$.

d Use the regression equation from part **c** and the seasonal indices to predict the takings for Monday, Wednesday and Saturday of week 4.

9 The number of students absent from a school on each day for the first four full weeks of a term are shown in the table below.

n	Week	Day	No. absent	5-pt MA (M)	Seasonally adjusted absences
1		Mon	35	–	32
2		Tue	25	–	
3	One	Wed	24	30.4	
4		Thur	28	30.8	
5		Fri	40	31.8	30
6		Mon	37	32.4	34
7		Tue	30	32.8	
8	Two	Wed	27	34.4	
9		Thur	30	36.2	
10		Fri	48	36.6	36
11		Mon	46	37.8	42
12		Tue	32	38.8	
13	Three	Wed	33	40.2	
14		Thur	35	40.6	
15		Fri	55	42.6	42
16		Mon	48	44.0	43
17		Tue	42	45.8	
18	Four	Wed	40	46.0	
19		Thur	44	–	
20		Fri	56	–	42

a Use a graphic calculator or computer spreadsheet to view the 5-point moving averages (M) on the vertical axis plotted against n on the horizontal axis, to confirm that for the variables n and M, linear regression would be appropriate.

b Use n and the 5-point moving averages, M, to determine the equation of the regression line $M = An + B$.

c Create a completed version of the following to calculate the seasonal indices.

	Monday	Tuesday	Wednesday	Thursday	Friday
Week 1	1.15132				1.31579
Week 2	1.07558				1.39535
Week 3	1.14428				1.36816
Week 4	1.04348				1.21739
Mean (4 dp)	1.1037				1.3242

d Calculate the missing values in the final column in the first table

e Use this regression equation $M = An + B$ and the seasonal indices to predict the absences for each day of week five.

f Suggest some real-life events that could make these predictions unreliable.

ISBN 9780170395069

10 Let us suppose that the number of rescues carried out by a surf rescue group in each two months of its six-months season, over four seasons are as follows:

n	Season	Months	Rescues	Season's 2-month mean
1		Oct/Nov	46	
2	1	Dec/Jan	111	82
3		Feb/Mar	89	
4		Oct/Nov	36	
5	2	Dec/Jan	104	76
6		Feb/Mar	88	
7		Oct/Nov	36	
8	3	Dec/Jan	107	70
9		Feb/Mar	67	
10		Oct/Nov	34	
11	4	Dec/Jan	91	64
12		Feb/Mar	67	

a Create a completed version of the following:

	Rescues as a percent of season's 2 month mean (as percentages and correct to 2 decimal places)		
	Oct/Nov	**Dec/Jan**	**Feb/Mar**
Season 1	56.10%	135.37%	
Season 2			
Season 3			
Season 4			
Seasonal indices (nearest percent)	52%		

b Produce deseasonalised figures, D, for each of the two-month periods of seasons 1 to 4. (Give the D values to the nearest integer.)

c Use a graphic calculator or computer spreadsheet to view a graph showing the deseasonalised figures, D, on the vertical axis plotted against n on the horizontal axis (n is as shown in the initial table), to confirm that for the variables n and D linear regression would be appropriate. Hence determine the equation of the regression line

$$D = A + Bn.$$

d Use the regression equation from part **b** and the seasonal indices to predict the number of rescues for each two-month period of season 5.

2. Moving averages and seasonal effects ●●●●●●●●●

Internet websites for government agencies such as

The Australian Bureau of Statistics
and
The Bureau of Meteorology

have a wealth of time series data available.

For example

Monthly numbers of house* building commencements approved.
(*Referred to as 'dwelling units'.)

Quarterly consumer sales figures.

Monthly maximum temperatures.

Etc.

Investigate.

Miscellaneous exercise two

This miscellaneous exercise may include questions involving the work of this chapter, the work of any previous chapters, and the ideas mentioned in the Preliminary work section at the beginning of the book.

1 Increase $250 by 10%, then decrease your answer by 10%.

2 The rule for converting a Celsius temperature (°C) to the corresponding Fahrenheit temperature (°F) is

$$F = \frac{9}{5}C + 32.$$

What is the value of the correlation coefficient r_{CF}?

3 What does it mean if the seasonal index for February is 0.87 (or 87%)?

4 Deseasonalise the following raw data given the seasonal indices are as stated.

Season	Spring	Summer	Autumn	Winter
Raw data	1400	1940	1520	750
Seasonal index	1.40	1.25	0.95	0.4

5

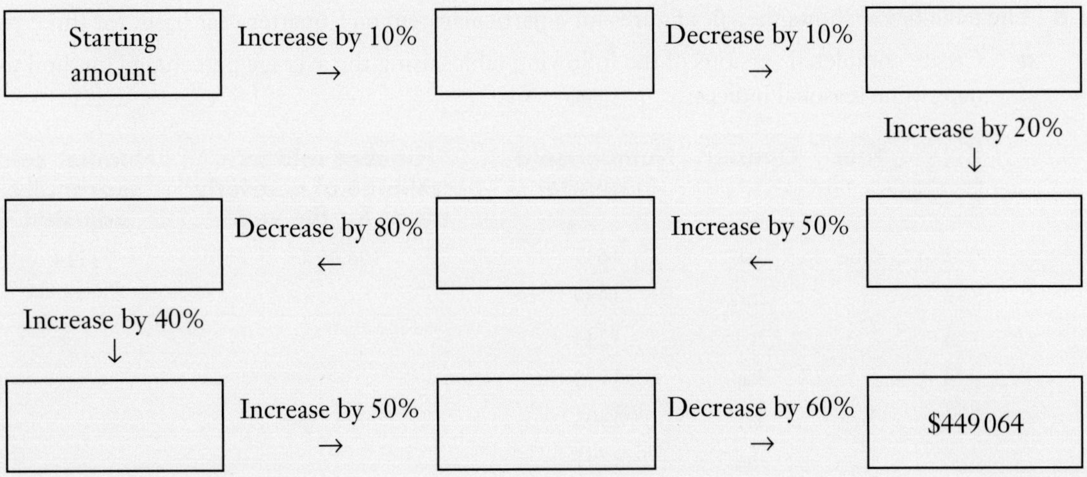

Starting amount	Increase by 10% →		Decrease by 10% →		

Increase by 20% ↓

	Decrease by 80% ←		Increase by 50% ←	

Increase by 40% ↓

	Increase by 50% →		Decrease by 60% →	$449 064

Find the starting amount.

6 Counting the year 1998 as zero, the year of manufacture and the asking price of a certain make and model of vehicle advertised in a newspaper one day in 2014 were as follows:

Year No.	5	13	6	4	4	15	6
Asking price ($)	19 980	44 500	17 000	13 500	11 400	57 300	20 500

Year No.	11	10	7	12	10	6	
Asking price ($)	39 000	32 900	25 000	35 700	36 800	25 500	

Display these figures graphically with t, the year number in the above table, on the horizontal axis and p, the asking price, on the vertical axis.

Comment on any relationships shown on your graph.

Use your calculator or a spreadsheet to determine the equation of the line of best fit in the form $p = at + b$ and use this to predict the likely price in 2014 for a 2007 vehicle of this make and model.

7 Using sales data from previous years a company determines that the quarterly revenues achieved from sales of its products seem to follow a reasonably steady seasonal pattern with the following seasonal indices:

1st quarter	0.85	2nd quarter	1.25
3rd quarter	1.12	4th quarter	0.78

The following year the company finds that the quarterly revenue figures are as follows:

1st quarter	$752 000	2nd quarter	$1 138 000
3rd quarter	$1 176 000	4th quarter	$694 000

a If the above information is all we know about the company, why are the quarterly revenues shown above a surprise?

b If we were informed that the company had a big advertising campaign at the start of the 3rd quarter might this make the revenue figures less surprising?

8 The table below shows the sales figures for a particular item on a quarter year basis, for three years.

a Create completed versions of the following tables using the average percentage method to determine seasonal indices.

t	Year	Quarter	Number sold in quarter	Number sold as percentage of quarterly mean for the year	Number sold seasonally adjusted
1	One	1st	1637	106.99%	1514
2		2nd	1489		
3		3rd	1244		
4		4th	1750		
5	Two	1st	1405		
6		2nd	1241		
7		3rd	1012		
8		4th	1538		
9	Three	1st	1253		
10		2nd	1121		
11		3rd	852		
12		4th	1362		

	Year 1	Year 2	Year 3	Seasonal index
1st quarter	106.99%			108.13%
2nd quarter				
3rd quarter				
4th quarter				

b Use a graphic calculator or computer to view the graph of the seasonally adjusted figures, which we will call S, plotted on the vertical axis, against t on the horizontal axis, and confirm that applying linear regression techniques is appropriate.

c Use t and the seasonally adjusted sales figures to determine the equation of the regression line $S = at + b$.

d Use this regression equation and the seasonal indices to predict the number of the particular item sold for each quarter of year 4.

ISBN 9780170395069

3.

Finance I: Saving and borrowing

Revision of simple interest, compound interest and recursive rules

The *Preliminary work* at the beginning of this book reminded you of simple interest, compound interest and sequences defined by giving the first term and a recursive rule.

The next examples and exercise 3A that follows should serve as a further reminder.

Note: The *simple interest*, I, earned when P is invested for T years in an account paying
$R\%$ per annum simple interest is given by $\qquad I = \dfrac{PRT}{100}$

If instead we use R in decimal form we use $\qquad I = PRT.$

For example if the interest rate is 6% the first formula would use $R = 6$ but the second would use $R = 0.06$.

WS

Simple interest

WS

Applications of simple interest

WS

Using spreadsheets to calculate interest

WS

Compound interest

EXAMPLE 1

How much interest would be earned in 4 years on an investment of $5000 at 6% per annum simple interest?

Solution

Using $\qquad I = PRT \qquad\qquad$ with $P = 5000$, $R = 0.06$ and $T = 4$
$$\begin{aligned} I &= 5000 \times 0.06 \times 4 \\ &= 1200 \end{aligned}$$

The interest earned will be $1200.

EXAMPLE 2

What annual rate of simple interest will see an investment of $8000 earn $3600 simple interest in 6 years?

Solution

Using $\qquad I = PRT \qquad\qquad$ with $I = 3600$, $P = 8000$ and $T = 6$
$$\begin{aligned} 3600 &= 8000 \times R \times 6 \\ 3600 &= 48\,000R \\ R &= \frac{3600}{48\,000} \\ &= 0.075 \end{aligned}$$

The required rate of simple interest is 7.5%.

EXAMPLE 3

A simple interest arrangement of 5.4% per annum sees an initial investment grow to $11 121.60 in 6 years. What was the initial investment?

Solution

Using

$$I = PRT \qquad \text{with } P + I = 11\,121.60,\ R = 0.054 \text{ and } T = 6$$

$$
\begin{aligned}
P + I &= P + PRT \\
11\,121.60 &= P + P \times 0.054 \times 6 \\
11\,121.60 &= P + 0.324P \\
11\,121.60 &= 1.324P \\
P &= 8400
\end{aligned}
$$

The initial investment was $8400.

EXAMPLE 4

$25 000 is invested for 4 years with an annual compound interest rate of 6%. Find the amount this account is worth at the end of the 4 years if the compounding occurs

a annually, **b** quarterly, **c** daily.

Solution

a Compounding annually.
There will be 4 compoundings in the 4 years, each of 6%.

$$
\begin{aligned}
\text{Value after 4 years} &= \$25\,000 \times 1.06 \times 1.06 \times 1.06 \times 1.06 \\
&= \$25\,000 \times 1.06^4 \\
&= \$31\,561.92 \text{ to the nearest cent.}
\end{aligned}
$$

b Compounding quarterly.
There will be 4 compoundings each year, each of 1.5% (= 6% ÷ 4).

I.e. 16 in the 4 years

$$
\begin{aligned}
\text{Value after 4 years} &= \$25\,000 \times 1.015^{16} \\
&= \$31\,724.64 \text{ to the nearest cent.}
\end{aligned}
$$

c Compounding daily.
There will be 365 compoundings each year, each of (6% ÷ 365).

I.e. 1460 (= 365 × 4) in the 4 years

$$
\begin{aligned}
\text{Value after 4 years} &= \$25\,000 \times \left(1 + \frac{0.06}{365}\right)^{365 \times 4} \\
&= \$31\,780.60 \text{ to the nearest cent.}
\end{aligned}
$$

Note:
- For convenience, in this book, we will ignore leap years and assume that all years have 365 days.
- Prior to the ready availability of computers and calculators, some banking calculations, again for convenience, were based on a concept called a *banker's year*. This concept takes a year as being 360 days and consisting of twelve equal months each of thirty days. This concept is mentioned here for information only, it will not be used in this course.

ISBN 9780170395069

EXAMPLE 5

At an 8% annual compound interest rate, with compounding every six months, how many years would it take for an initial investment of $2500 to grow to $3700?

Solution

There will be 2 compoundings each year, each of 4%.

If the account runs for T years:

$$\$3700 \; = \; \$2500 \times 1.04^{2T}$$

Using the solve facility of some calculators gives $T = 4.998$ (rounded to 3 decimal places).

It would take 5 years.

Consider an investment of $8000 at 5% per annum simple interest. The annual value of this investment forms the sequence:

$$\$8000 \qquad \$8400 \qquad \$8800 \qquad \$9200 \qquad \$9600 \qquad \dots$$

i.e. an arithmetic progression with first term $8000 and common difference $400.

Recursive rule: $\qquad\qquad\qquad T_1 = \$8000, \quad T_{n+1} = T_n + \$400.$

Or, if we want T_n to represent the value after n years, $T_0 = \$8000, \quad T_{n+1} = T_n + \$400.$

Had the investment been at 5% compound interest, with interest compounded annually, the sequence would be:

$$\$8000 \qquad \$8400 \qquad \$8820 \qquad \$9261 \qquad \$9724.05 \qquad \dots$$

i.e. a geometric progression with first term $8000 and common ratio 1.05.

Recursive rule: $\qquad\qquad\qquad T_1 = \$8000, \quad T_{n+1} = 1.05 \times T_n.$

Or, if we want T_n to represent the value after n years, $T_0 = \$8000, \quad T_{n+1} = 1.05 \times T_n.$

In this way recursive rules can be useful for simple and compound interest situations.

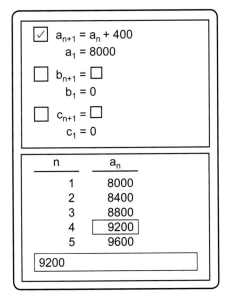

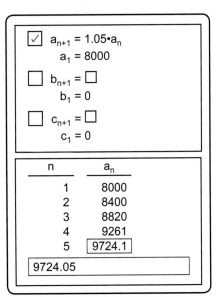

Exercise 3A

Simple interest

1 What will be the value after 6 years of $2000 invested at 6% per annum simple interest?

2 How long will it take for $800, invested in an account paying 7.5% per annum simple interest, to become $1100?

3 What annual rate of simple interest is needed to see an investment of $5000 become $7500 in 4 years?

4 How much must have been invested in an account paying 8% simple interest if after 6 years the total interest earned was $1152?

5 What annual rate of simple interest is needed to see an initial investment of $6600 become $7007 in 10 months?

6 A simple interest arrangement of 6.5% per annum sees an initial investment grow to $10 703 in 6 years. What was the initial investment?

Compound interest

7 $4000 is invested at 8% per annum compounded annually. Determine the value of this investment after 6 years.

8 $500 is left in an account paying 6% per annum compounded six-monthly. How much will this account be worth 50 years later?

9 How much interest is earned on $50 000 invested for 4 years if the interest rate is 6% per annum compounded annually?

How much more interest would be earned in the 4 years if instead the interest is compounded monthly?

10 How much interest is earned on $400 000 invested for ten years if the interest rate is 8% per annum compounded annually?

How much more interest would be earned in the ten years if instead the interest is compounded daily?

11 How much must be invested now into an account paying 7.5% compound interest, compounded annually, for the account to be worth $10 000 in five years?

12 If an amount is invested at 6% per annum compound interest, with interest compounded quarterly, how long would it take for the initial investment to double in value?

13 At a 9% annual compound interest rate, with compounding every six months, how long would it take for an initial investment of $5000 to first exceed $12 000?

ISBN 9780170395069

Recursive rules

Determine the first five terms and, with the help of a calculator that is able to handle recursive rules if necessary, the fifteenth term, of each of the following recursively defined sequences.

14 $T_{n+1} = T_n + 3$, $\qquad\qquad$ $T_1 = 5$.

15 $T_{n+1} = T_n - 3$, $\qquad\qquad$ $T_1 = 5$.

16 $u_{n+1} = u_n + 5$, $\qquad\qquad$ $u_1 = -10$.

17 $\quad a_n = a_{n-1} + 2.5$, $\qquad\quad$ $a_1 = 12.5$.

18 $T_{n+1} = 2 \times T_n$, $\qquad\qquad$ $T_1 = 0.25$.

19 $T_{n+1} = 1.5 \times T_n$, $\qquad\qquad$ $T_1 = 20\,480$.

20 $a_{n+1} = 0.5 \times a_n$, $\qquad\qquad$ $a_1 = 2\,621\,440$.

21 $u_{n+1} = 0.5 \times u_n$, $\qquad\qquad$ $u_1 = 2^{30}$. $\qquad\quad$ (Give answers as powers of 2.)

22 $T_{n+1} = 2 \times T_n + 3$, $\qquad\qquad$ $T_1 = 5$.

23 $T_{n+1} = 1.2 \times T_n + \$100$. $\qquad$ $T_1 = \$1000$. $\qquad$ (Write T_{15} to the nearest cent.)

24 $T_{n+1} = 1.05 \times T_n - \$250$. $\qquad$ $T_1 = \$5000$.

25 $\quad T_n = a_n + u_n \qquad$ where $\qquad a_{n+1} = a_n + 6$, $\qquad a_1 = 5$,
$\qquad\qquad\qquad\qquad$ and $\qquad u_{n+1} = u_n + 5$, $\qquad u_1 = 7$.

26 $\quad T_n = a_n \times u_n \qquad$ where $\qquad a_{n+1} = 2 \times a_n$, $\qquad a_1 = 5$,
$\qquad\qquad\qquad\qquad$ and $\qquad u_{n+1} = 0.5 \times u_n$, $\qquad u_1 = 163\,840$.

27 $T_{n+2} = T_{n+1} + T_n$, $\qquad\qquad$ $T_1 = 1$, $\qquad\qquad$ $T_2 = 1$.

28 $T_{n+2} = 2 \times T_{n+1} + T_n$, $\qquad\quad$ $T_1 = 2$, $\qquad\qquad$ $T_2 = 3$.

29 $T_{n+1} = (-1)^n T_n$, $\qquad\qquad$ $T_1 = 5$.

30 $T_{n+1} = (-1)^n \times T_n + 2$, $\qquad\quad$ $T_1 = 5$.

Effective annual interest rate

Consider $100 invested in an account paying 6% per annum compound interest.

After one year, with compounding …

occurring annually, the account will be worth

$$\$100 \times 1.06$$
$$= \$106$$

occurring six monthly, the account will be worth

$$\$100 \times 1.03^2$$
$$= \$106.09$$

occurring quarterly, the account will be worth

$$\$100 \times 1.015^4$$
$$= \$106.136 \text{ (correct to 3 decimal places)}$$

occurring monthly, the account will be worth

$$\$100 \times 1.005^{12}$$
$$= \$106.168 \text{ (correct to 3 decimal places)}$$

occurring daily, the account will be worth

$$\$100 \times \left(1 + \frac{0.06}{365}\right)^{365}$$
$$= \$106.183 \text{ (correct to 3 decimal places)}$$

Thus whilst 6% per annum returns $6 interest in the year when compounded annually, more frequent compoundings sees our 6% interest rate return more than $6 on $100 invested, i.e. more than 6%. The 6% is said to be the **nominal annual interest rate** but to make comparison between the effect of the different compounding periods we call the 6.09%, 6.136%, 6.168% and 6.183% returns shown above the **effective annual interest rates**.

Thus for a nominal compound interest rate of 6% per annum the effective annual interest rate for six-monthly compounding is 6.09%. The effective annual interest rate for quarterly compounding is 6.136% (correct to three decimal places), the effective annual interest rate for monthly compounding is 6.168% (correct to three decimal places) etc.

This effective annual interest rate can be determined by consideration of the compound interest earned in one year by an investment of $100, as shown above, or by use of the following formula:

For a nominal interest rate of i per annum and n compounding periods per year,

$$\text{Effective annual interest rate} = \left(1 + \frac{i}{n}\right)^n - 1.$$

This is sometimes written as

$$i_{\text{effective}} = \left(1 + \frac{i}{n}\right)^n - 1.$$

Readers should check that with suitable rounding this formula does indeed give 0.0609 (i.e. 6.09%), 0.061 36 (i.e. 6.136%) and 0.061 68 (i.e. 6.168%), for $i = 0.06$ and values for n of 2, 4 and 12 respectively, as quoted above.

EXAMPLE 6

Find the effective annual interest rate for a nominal interest rate of 8% per annum with compounding occurring quarterly.

Solution

By considering $100.

8% per annum is 2% per quarter.

After 1 year a $100 investment becomes

$$\$100 \times 1.02^4$$

$$= \$108.243 \text{ (correct to 3 decimal places)}$$

By the formula.

Using $i_{\text{effective}} = \left(1 + \dfrac{i}{n}\right)^n - 1$

with $i = 0.08$

and $n = 4$.

$$i_{\text{effective}} = \left(1 + \dfrac{0.08}{4}\right)^4 - 1$$

$$= 0.082\,432\,16$$

The effective annual interest rate is 8.243% (correct to 3 decimal places).

Note: If a question simply refers to an annual interest rate of, say, 8%, it should be assumed that it is the nominal annual rate that is being quoted.

Exercise 3B

1 Copy and complete the following table. (1 year = 12 months = 52 weeks = 365 days.)

Compounding frequency	Nominal annual interest rate (%)	Effective annual interest rate (%)
Annual	4%	
Six-monthly	4%	
Quarterly	4%	
Monthly	4%	
Weekly	4%	
Daily	4%	

2 Copy and complete the following table. (1 year = 12 months = 52 weeks = 365 days.)

Compounding frequency	Nominal annual interest rate (%)	Effective annual interest rate (%)
Annual	8%	
Six-monthly	8%	
Quarterly	8%	
Monthly	8%	
Weekly	8%	
Daily	8%	

3 Looking at your answers for the effective annual interest rates in questions **1** and **2**, are the rates for the 8% situation simply twice those of the 4% situation?

If not, why not?

Initial deposit plus regular investments

The investments considered so far in this chapter have all involved a 'one-off' investment deposited in an account and left to earn interest until some later date. In practice, many savings schemes involve an initial amount being invested followed by further deposits into the account, perhaps on a regular basis. Consider for example an initial investment of $5000 into an account paying compound interest of 6% per annum, compounded annually, with an extra $500 deposited into the account each year thereafter.

$$
\begin{aligned}
\text{Initial deposit} \quad &= \quad \$5000 \\
\text{Amount in account at end of 1 year} \quad &= \quad \$5000 \times 1.06 + \$500 \\
&= \quad \$5800 \\
\text{Amount in account at end of 2 years} \quad &= \quad \$5800 \times 1.06 + \$500 \\
&= \quad \$6648 \\
\text{Amount in account at end of 3 years} \quad &= \quad \$6648 \times 1.06 + \$500 \\
&= \quad \$7546.88
\end{aligned}
$$

With $T_1 = \$5000$, $T_2 = \$5800$, $T_3 = \$6648$, $T_4 = \$7546.88$ etc, what we have here is a sequence that can be defined recursively by the rule:

$$T_{n+1} \quad = \quad 1.06 \times T_n + \$500. \qquad T_1 \quad = \quad \$5000.$$

Or, if we want the end of year number to be the same as the term number we use

$$T_{n+1} \quad = \quad 1.06 \times T_n + \$500. \qquad T_0 \quad = \quad \$5000.$$

Then T_1 will give the value at the end of year one, T_2 will give the value at the end of year two, T_3 will give the value at the end of year three, etc.

The progress of the investment can then be easily viewed using a calculator with an ability to display the terms of a recursively defined sequence, or using a computer spreadsheet – see the displays on the next page.

Calculator display (above left):

- ☑ $a_{n+1} = 1.06 \cdot a_n + 500$
  $a_1 = 5000$
- ☐ $b_{n+1} = \square$
  $b_1 = 0$
- ☐ $c_{n+1} = \square$
  $c_1 = 0$

n	a_n
4	7546.9
5	8499.7
6	9509.7
7	10580.
8	11715.

11715.0701198851

Spreadsheet (above right):

	A	B	C	D
1	Initial investment			$5,000.00
2	Interest rate per yr (%)			6.00
3	Annual deposit			$500.00
4	Balance at end of year		1	$5,800.00
5			2	$6,648.00
6			3	$7,546.88
7			4	$8,499.69
8			5	$9,509.67
9			6	$10,580.25
10			7	$11,715.07
11			8	$12,917.97
12			9	$14,193.05
13			10	$15,544.64

- Remember that in the display above left, with the first term being the value after zero years, the 8th term will be the value of the investment after 7 years.

Create a spreadsheet like the one shown above right. Try to make your spreadsheet adaptable to other situations of this type such that by simply changing the amounts in cells D1, D2 and D3 the annual balances are automatically recalculated.

EXAMPLE 7

At the beginning of one month, $1000 is invested into an account paying interest at 7.5% per annum, compounded monthly, and an extra $100 is invested at the end of that first month and the end of every month thereafter. How much is the account worth at the end of the month two years later, just after the $100 deposit for that month is made? (For the purposes of this question assume that a year consists of 12 equal months.)

Solution

Initial value $= \$1000$

Value at end of 1st month $= \$1000 \times \left(1 + \dfrac{0.075}{12}\right) + \$100$ i.e. $\$1106.25$

Value at end of 2nd month $= \$1106.25 \times 1.00625 + \$100$

Thus with T_n the value of the investment at the end of the nth month

$$T_{n+1} = 1.00625 \times T_n + \$100. \qquad T_0 = \$1000.$$

Either using the facility of some calculators to display the terms of recursively defined sequences, or by using your version of the spreadsheet shown above right, with appropriate entries suitably changed (see next page):

Value at end of 24th month $= T_{24}$
$= \$3741.96$ (nearest cent).

Hence two years after the initial investment, and just after the $100 deposit for that month has been made, the account will be worth $3741.96.

	A	B	C	D
1	Initial investment			$1,000.00
2	Interest rate per month (%)			0.625
3	Monthly deposit			$100.00
4	Balance at end of month		1	$1,106.25
5			2	$1,213.16
6			3	$1,320.75
7			4	$1,429.00
8			5	$1,537.93
9			6	$1,647.54
10			7	$1,757.84
11			8	$1,868.83
12			9	$1,980.51
13			10	$2,092.89
14			11	$2,205.97
15			12	$2,319.75
16			13	$2,434.25
17			14	$2,549.47
18			15	$2,665.40
19			16	$2,782.06
20			17	$2,899.45
21			18	$3,017.57
22			19	$3,136.43
23			20	$3,256.03
24			21	$3,376.38
25			22	$3,497.48
26			23	$3,619.34
27			24	$3,741.96

☑ $a_{n+1} = 1.00625 \cdot a_n + 100$
$a_0 = 1000$

☐ $b_{n+1} = \square$
$b_0 = 0$

☐ $c_{n+1} = \square$
$c_0 = 0$

n	a_n
20	3256.0
21	3376.4
22	3497.5
23	3619.3
24	3742.0

3741.9643079925

Some calculators have built-in financial programs that allow some standard financial calculations to be performed automatically, given appropriate information.

The display on the right shows typical information that would be displayed for the situation just encountered.

The N in the display is the number of investment periods, in this case 24 months.

I (%) is the annual interest rate as a percentage, in this case 7.5.

The initial investment of $1000 is shown as PV, for present value. This is shown as a negative amount because from the investor's point of view it is money being deposited (outgoing), not money being received.

The additional investment per month (Pmt) is $100, and again from the investor's point of view this is shown as a negative (outgoing).

Payments per year and compoundings per year are each 12.

Provided the program has the correct settings, e.g. the payment date set as being at the end of each compounding period, not at the beginning, when asked to determine the final value (FV) the calculator returns the final value as $3741.96, as before.

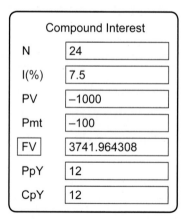

Compound Interest

N	24
I(%)	7.5
PV	−1000
Pmt	−100
FV	3741.964308
PpY	12
CpY	12

ISBN 9780170395069

Exercise 3C

For questions **1** and **2** use recursion. For **3, 4** and **5** do each question *twice*, once using recursion and once using a calculator with financial capability.

1 $4000 is invested into an account paying interest at 8% per annum, compounded annually, and an extra $200 is invested at the end of each 12 months. Thus:

Amount in account at end of 1 year	=	$4000 × 1.08 + $200	← T_1
Amount in account at end of 2 years	=	($4000 × 1.08 + $200) × 1.08 + $200	← T_2

Express T_{n+1} in terms of T_n and determine (to the nearest cent) the amount in the account at the end of ten years, after the $200 for that year has been added.

2 $5000 is used to open an investment account paying interest at 6% per annum, compounded monthly, and an extra $100 is invested at the end of each one-month period thereafter. Thus, the value of the account progresses as follows:

Initially	=	$5000	← T_0
At the end of 1 month	=	$5000 × 1.005 + $100	← T_1
At the end of 2 months	=	($5000 × 1.005 + $100) × 1.005 + $100	← T_2

a Express T_{n+1} in terms of T_n.

b The account is closed at the end of three years, without the $100 for the end of the final month being made. Determine the final value of the account.

3 At the beginning of 2014 Tenielle opens a savings account by depositing $2500 into an account paying 8% per annum, compounded annually. Tenielle plans to follow that initial deposit with further deposits each of $1000 made at the end of every period of one year thereafter.

Assuming Tenielle keeps to her plans with regards to the further deposits, how much is the account worth at the beginning of 2025 (i.e. just after the $1000 deposit for the end of 2024 has been made)?

4 At the beginning of 2014 Sanchez opens a savings account by depositing $500 into an account paying 7.8% per annum, compounded annually. Sanchez plans to follow that initial deposit with further deposits each of $500 made at the end of every period of one year thereafter.

Assuming Sanchez keeps to his plans with regards to the further deposits, how much is the account worth at the end of 2020, i.e. just before the $500 deposit is made?

5 At the beginning of one month $5000 is invested into an account paying interest at 9% per annum, compounded monthly, and an extra $200 is invested at the end of that month and the end of every month thereafter. How much is the account worth at the end of the month three years later, just after the $200 deposit for that month?

6 $5000 is used to open an account paying interest that is compounded annually, and an additional $1000 is to be added to the account at the end of every one-year period thereafter. Using a spreadsheet, or a calculator able to display the terms of a recursively defined sequence, or financial programs available on some calculators or internet sites, find the annual interest rate required for this account to be worth

a $10000 after 3 years (including the final $1000 payment),

b $100000 after 25 years (including the final $1000 payment).

Compounding isn't only about investing

In a financial context, compounding is not only applied when we invest money, it is also applied to the interest we owe on borrowed money. For example, suppose you borrow $50 000 from a bank or financial institution. You will be charged interest on this loan. Let us suppose that the interest is charged at the rate of 6% per annum nominal rate, compounded annually.

After one year you will owe	$50 000 \times 1.06$	i.e. $53 000
After two years you will owe	$50 000 \times 1.06^2$	i.e. $56 180
After three years you will owe	$50 000 \times 1.06^3$	i.e. $59 550.80 etc.

If instead the compounding occurred monthly:

After one year you will owe	$50 000 \times 1.005^{12}$	i.e. $53 083.89
After two years you will owe	$50 000 \times 1.005^{24}$	i.e. $56 357.99
After three years you will owe	$50 000 \times 1.005^{36}$	i.e. $59 834.03 etc.

Similar repeated multiplication on an annual basis can occur when the value of an asset is **depreciated** each year.

Suppose a company purchases a new machine worth $100 000. One year later the machine is no longer new so its value will have depreciated. One method used for determining the value at the end of each year is to depreciate the value of the asset by a certain percentage of its value at the beginning of that year. Thus if the machine initially worth $100 000 is depreciated at an annual rate of 12% its value in later years will be as follows:

Value of the machine after 1 year	=	$100 000 \times 0.88$	i.e. $88 000	
Value of the machine after 2 years	=	$100 000 \times 0.88^2$	i.e. $77 000	(nearest $1000)
Value of the machine after 3 years	=	$100 000 \times 0.88^3$	i.e. $68 000	(nearest $1000)
		etc.		

You may recall from Unit 3 of *Mathematics Applications* that the above form of depreciation, called the **reducing balance** method (or fixed percentage method or diminishing value method) is not the only method for calculating depreciation.

We also have: the **flat rate**, or straight line, method, in which the depreciation is a fixed amount each year,

and the **unit cost** method, in which the depreciation is related to the units of production the machine has produced.

For example suppose a machine has an initial value of $500 000.

Consider the following three depreciation schedules:

Flat rate depreciation of $50 000 per year

Initial value:	$500 000		
Value after 1 year:	$450 000	Depreciation in 1st year:	$50 000
Value after 2 years:	$400 000	Depreciation in 2nd year:	$50 000
Value after 3 years:	$350 000	Depreciation in 3rd year:	$50 000
Value after 4 years:	$300 000	Depreciation in 4th year:	$50 000 etc.

Reducing balance depreciation of 10% per year

Initial value:	$500 000		
Value after 1 year:	$500 000 × 0.9 = $450 000	Depreciation in 1st year:	$50 000
Value after 2 years:	$450 000 × 0.9 = $405 000	Depreciation in 2nd year:	$45 000
Value after 3 years:	$405 000 × 0.9 = $364 500	Depreciation in 3rd year:	$40 500
Value after 4 years:	$364 500 × 0.9 = $328 050	Depreciation in 4th year:	$36 450 etc.

Unit cost depreciation of $60 000 per 100 000 units of production

Initial value:	$500 000
Value after 100 000 units of production:	$440 000
Value after 200 000 units of production:	$380 000
Value after 300 000 units of production:	$320 000 etc.

Exercise 3D

1 An asset has an initial value of $220 000. Determine its value at the end of each of the first three years using **a** flat rate depreciation of $25 000 per year,

b reducing balance depreciation of 15% per year.

2 An asset has an initial value of $80 000. By how much does the asset depreciate in each of its first three years if the depreciation method used is **a** flat rate depreciation of $10 000 per year?

b reducing balance depreciation of 20% per year?

3 A printing machine initially worth $30 000 is expected to be worth just $2000 after it has been used to produce one million copies.

Using the unit cost depreciation method what will be the value of the machine after it has produced

a 250 000 copies? **b** 500 000 copies? **c** 750 000 copies?

4 Some calculators with financial capabilities, and some online financial calculators, can work out depreciation schedules. Investigate.

Loans with regular repayments

Rather like the investments that involved a regular contribution, people often make regular contributions to reduce the balance of a loan.

For example, suppose Jim wants to buy a motor scooter costing $4000.

Further suppose that Jim borrows the $4000 from a bank, that he is charged 1% interest per month on the remaining balance and he manages to repay $100 at the end of every period of one month.

Initially he owes	$4000		
After 1 month he will owe:	$4000 × 1.01 − $100	=	$3940
After 2 months he will owe:	$3940 × 1.01 − $100	=	$3879.40
After 3 months he will owe:	$3879.40 × 1.01 − $100	=	$3818.194 etc.,

the value of the loan reducing each month because the amount repaid exceeds the interest charged. Such an arrangement is referred to as a **reducing balance loan**.

EXAMPLE 8

Johan takes out a loan of $7000. Interest of 9.4% per annum is added to the loan annually and Johan repays $800 at the end of each one-year period. How much will Johan still owe immediately after he makes the $800 repayment at the end of year 8?

Solution

Initial amount owed:	$7000	
Amount owed after 1 year:	$7000 \times 1.094 - $800	= $6858
Amount owed after 2 years:	$6858 \times 1.094 - $800	= $6702.65 (nearest cent)
Amount owed after 3 years:	$6702.65 \times 1.094 - $800	= $6532.70 (nearest cent)

We have $T_{n+1} = T_n \times 1.094 - \$800$,

and $T_0 = \$7000$

Using the recursion capability of some calculators, or a spreadsheet:

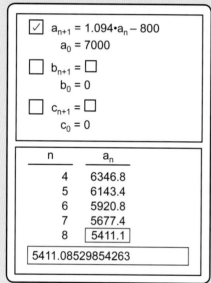

	A	B	C	D
1	Amount borrowed			$7,000.00
2	Interest rate			9.40
3	Annual repayments			$800.00
4	Balance at end of year		1	$6,858.00
5			2	$6,702.65
6			3	$6,532.70
7			4	$6,346.78
8			5	$6,143.37
9			6	$5,920.85
10			7	$5,677.41
11			8	$5,411.09
12				
13				

The amount still owed at the end of year 8 is $5411.09.

Note:
- Continuing down either of the previous tables allows us to see when the loan will be reduced to zero. In this case the payment at the end of the twentieth year would pay off the loan (and this final payment would only need to be approximately $200, not $800).

n	a_n
16	2150.9
17	1553.1
18	899.10
19	183.62
20	−599.1

−599.121602634385

- By redefining the recursive rule using a different repayment amount, or by inserting a different repayment amount into the appropriate cell of the spreadsheet, we can investigate how much the annual repayment needs to be in order to pay off the loan in the eight years. Investigate this for yourself and confirm that in this case the required amount is $1283.58 (with the final repayment being a few cents more).

Once again we could use the financial programs available on some calculators and internet websites that allow the remaining balance of a loan after a number of repayments, or the repayment required to pay off a loan in a particular time etc., to be determined. The display on the right shows the $1283.58 mentioned in the previous dot point (i.e. the regular repayment needed to pay the loan off in 8 years.) Whilst these programs can be useful make sure that if the course requires you to be able to use spreadsheets or recursive formulae you can do so.

The display shows that repayments (*PMT*) of $1283.584243, i.e. $1283.58 to the nearest cent, made once per year (*P/Y*), will see a loan of $7000 reduced to a future value (*FV*) of $0 in 8 years when compound interest is charged at 9.4% per annum, compounded annually, i.e. 1 compounding per year (*C/Y*). In the display the repayment is shown as a negative – it is money being *taken from us* and put into the account. The $7000 on the other hand is shown as positive – it is money being *given to us* in the form of a loan.

(Paying $1283.58 involves rounding the exact payment *down*, hence the final payment will be a few cents more than $1283.58.)

Compound Interest	
N	8
I(%)	9.4
PV	7000
PMT	−1283.584243
FV	0
P/Y	1
C/Y	1

Using finance solvers for compound interest loans

Exercise 3E

1 Use the display on the right to determine what goes in the blank spaces A to I in the following:

Repayments of __A__ , i.e. __B__ to the nearest cent, made __C__ times per year, will see a loan of __D__ reduced to a future value of __E__ in __F__ years when compound interest is charged at __G%__ per year, compounded __H__ , that is, __I__ compoundings per year.

Compound Interest	
N	24
I(%)	9
PV	4000
Pmt	−182.7389691
FV	0
PpY	12
CpY	12

2 At the beginning of one year Jane borrows $10 000. She repays $2000 at the end of that year and at the end of each year thereafter (except the last) until the loan is paid off. Interest is calculated at 10% per annum, compounded annually. The table on the right shows the progress of the loan (to the nearest cent).

Determine A, B, C and D.

Year	Amount owed at beginning of year	Amount owed at end of year
1	$10 000	$10 000 × 1.1 − $2000 = $9000
2	$9000	$9000 × 1.1 − $2000 = $7900
3	$7900	$7900 × 1.1 − $2000 = $6690
4	$6690	$6690 × 1.1 − $2000 = $A
5	$A	$A × 1.1 − $2000 = $B
6	$B	$B × 1.1 − $2000 = $2284.39
7	$2284.39	$2284.39 × 1.1 − $2000 = $C
8	$C	$C × 1.1 − $D = $0

3 Ranii borrows $1600 at the beginning of one month and repays $300 at the end of that month, and every month thereafter (except the last) until the loan is repaid.

Interest is calculated at 1.5% per month.

The table below shows the progress of the loan (to the nearest cent).

Month	Amount owed at beginning of month	Amount owed at end of month
1	$1600	$1600 \times A - $300 = $1324
2	$1324	$1324 \times A - $300 = $B
3	$B	$B \times A - $300 = $759.52
4	$759.52	$759.52 \times A - $300 = $470.91
5	$470.91	$470.91 \times A - $300 = $177.97
6	$177.97	$177.97 \times A - $C = $0

Determine the values of A, B and C.

Use recursion techniques for questions **4**, **5** and **6**.

4 If $A is borrowed at a monthly interest rate of 1%, with $P being repaid at the end of each month, the amount owing after $(n + 1)$ months is related to the amount owing after n months according to the rule

$$T_{n+1} = 1.01(T_n) - P, \quad \text{for } n \geq 0,$$

where $T_0 = A$,

and T_n is the amount owing after n months.

Find the amount owing at the end of each of the first four months if $50 000 is borrowed at a monthly interest rate of 1% and $800 is repaid at the end of each month.

5 Hisham takes out a loan for $24 000 agreeing to repay $350 at the end of each monthly period, after monthly interest is added at 1% of the outstanding balance. How much will he still owe on this loan after 5 years?

Suppose instead that Hisham had made monthly repayments of $375. How much would he still owe after 5 years in this case?

6 Jackson needs to borrow some money from a finance company so that he can purchase a car costing $11 600. He is considering three different repayment schemes, all of which involve him in making regular monthly repayments and with interest added at the rate of 15% per year (i.e. 1.25% per month) calculated monthly on the outstanding balance before the repayment is made.

Scheme A: Borrow the full $11 600 and make monthly repayments of $250.
Scheme B: Borrow the full $11 600 and make monthly repayments of $290.
Scheme C: Pay $1200 that he has saved towards the cost of the car, borrow the rest and make monthly repayments of $290.

How long will each scheme take to pay off the loan?

ISBN 9780170395069

Use the inbuilt financial programs available on some calculators or a suitable online financial calculator to answer each of questions **7**, **8** and **9**.

7 To the nearest cent what constant monthly amount needs to be repaid to see a loan of $10 000, with compound interest of 6% per annum compounded monthly, paid off in 2 years?

8 To the nearest cent what constant monthly amount needs to be repaid to see a loan of $50 000, with compound interest of 12% per annum compounded monthly, paid off in 10 years?

9 To the nearest cent what constant monthly amount needs to be repaid to see a loan of $235 000, with compound interest of 9% per annum compounded monthly, paid off in 25 years?

10 The spreadsheet below left is for the situation of borrowing $4000, paying interest of 1% per month and repaying $100 per month. The spreadsheet below right shows the same situation but now with $150 repaid each month.

	A	B	C	D
1	Amount borrowed			**$4,000.00**
2	Interest rate per month			**1.00%**
3	Monthly repayments			**$100.00**
4	Balance at end of month		1	$3,940.00
5			2	$3,879.40
6			3	$3,818.19
7			4	$3,756.38
8			5	$3,693.94
9			6	$3,630.88
10			7	$3,567.19
11			8	$3,502.86
12			9	$3,437.89
13			10	$3,372.27
14			11	$3,305.99
15			12	$3,239.05
16			13	$3,171.44
17			14	$3,103.15
18			15	$3,034.19
19			16	$2,964.53
20			17	$2,894.17
21			18	$2,823.12
22			19	$2,751.35
23			20	$2,678.86
24			21	$2,605.65
25			22	$2,531.70
26			23	$2,457.02
27			24	$2,381.59

	A	B	C	D
1	Amount borrowed			**$4,000.00**
2	Interest rate per month			**1.00%**
3	Monthly repayments			**$150.00**
4	Balance at end of month		1	$3,890.00
5			2	$3,778.90
6			3	$3,666.69
7			4	$3,553.36
8			5	$3,438.89
9			6	$3,323.28
10			7	$3,206.51
11			8	$3,088.58
12			9	$2,969.46
13			10	$2,849.16
14			11	$2,727.65
15			12	$2,604.92
16			13	$2,480.97
17			14	$2,355.78
18			15	$2,229.34
19			16	$2,101.63
20			17	$1,972.65
21			18	$1,842.38
22			19	$1,710.80
23			20	$1,577.91
24			21	$1,443.69
25			22	$1,308.13
26			23	$1,171.21
27			24	$1,032.92

Create such a spreadsheet yourself, or adapt one you have already created, and find how much needs to repaid per month in order to pay the loan off in exactly two years.

Suppose instead the interest rate was 1.2% per month. How much would need to be repaid per month now to repay the loan in exactly two years?

3. Finance I: Saving and borrowing ●●●◦◦◦◦◦

What price house can these people afford?

Let us suppose that in order to purchase a house costing $480 000 John and Maxine pay a deposit of $30 000 from their savings and borrow the remaining $450 000 from a financial institution. The institution charges a fixed 8% per annum compound interest with compounding occurring monthly. John and Maxine arrange to repay the loan by making regular monthly payments for twenty-five years.

Using a calculator with the ability to perform financial calculations, or a similar financial capability on a computer or online, we can determine that their monthly repayments would be $3473.17, to the nearest cent.

Compound Interest	
N	300
I(%)	8
PV	450000
PMT	−3473.172987
FV	0
P/Y	12
C/Y	12

Before entering into such an arrangement John and Maxine would probably have first worked out what they could afford to pay per month.

Let us suppose that Kym and Hasim have $35 000 saved to put down as a deposit on a house. They feel they can afford to pay $2000 per month towards the cost of a loan. They plan to get a loan for a term of 25 years making regular monthly payments and are told the interest rate is 7.6% per annum compounded monthly.

What value house can they afford?

As the display on the right indicates, Kym and Hasim can afford a loan of $268 000, to the nearest $1000.

Therefore, with their deposit of $35 000, they could afford a house costing $303 000 (= $268 000 + $35 000).

Compound Interest	
N	300
I(%)	7.6
PV	268273.4425
PMT	−2000
FV	0
P/Y	12
C/Y	12

ISBN 9780170395069

In the case of Kym and Hasim if they borrow \$268 000 they will make 300 repayments each of \$1997.96 (see the display on the right).

Hence their total repayments will be

$$300 \times \$1997.96 \ = \ \$599\,388.$$

Thus on their loan of \$268 000 they have paid

$$\$599\,388 - \$268\,000 \ = \ \$331\,388 \text{ interest,}$$

… or have they? Read on.

By switching to '*Amortization*', as shown below right, some calculators show this total interest amount, and other information about the loan. Some calculators, and some online amortization calculators, can display a table showing month-by-month details about the loan.

Wanting the total interest paid from payment 1 to payment 300 the display on the right shows *PM1* as 1 and *PM2* as 300 and displays ΣINT, as shown.

If we round the ΣINT amount shown in the display to the nearest dollar, we again get \$331 388.

Wondering where the extra \$0.4392 came from?

Well, one calculation used 300 lots of \$1997.96 and the other used 300 lots of \$1997.961 464 ….

Explore your calculator with regard to amortization and investigate *online* amortization calculators.

Compound Interest	
N	300
I(%)	7.6
PV	268000
PMT	−1997.961464
FV	0
P/Y	12
C/Y	12

Amortization	
PM1	1
PM2	300
I%	7.6
PV	268000
PMT	−1997.961464
P/Y	12
C/Y	12
BAL	
INT	
PRN	
ΣINT	−331388.4392
ΣPRN	

- The slight rounding down of \$1997.961 464 to \$1997.96 means that in theory there would still be a very small amount (\$1.3051) remaining on the loan after the 300th payment. In practice, to clear the loan on the 300th payment, the final payment would be \$1997.96 + \$1.31, i.e. \$1999.27.

  The total interest amount of \$331 388
  (= 300 × \$1997.96 − \$268 000) quoted at the top of the page has not allowed for this change in the final payment. Thus a more accurate calculation of the total interest would be

  $$\$331\,389.31 \ (= 299 \times \$1997.96 + 1 \times \$1999.27 - \$268\,000),$$

  a total directly obtainable from ΣINT in the amortization calculation with *PM1* = 1, *PM2* = 300 and monthly repayment of \$1997.96.

  However, the \$331 388 would often be considered a satisfactory approximation.

- Under different rounding regimes slight variations are possible. If we round each month to the nearest cent, as some online amortization calculators do, the final payment becomes \$1999.06 and the total interest is then

  $$\$331\,389.10 \ (= 299 \times \$1997.96 + 1 \times \$1999.06 - \$268\,000).$$

  If you do ever think of taking out a loan to buy a house you should seek expert advice specific to your situation. The work here just introduces the mathematics.

- Amortization can also be written amortisation. However, we will use the z spelling here as that is how your calculator is likely to show it.

Exercise 3F

What price house can each of the following people afford? In each case assume that the interest rate shown is per annum, compounded monthly.

		Amount they can afford to repay per month	Deposit saved	Duration of loan	Interest rate
1	Fran and Michael	$3800	$17000	25 years	7.5%
2	Peta and Peter	$2400	$20000	25 years	8.2%
3	Rania and Umar	$4000	$35000	20 years	9.1%
4	Hue and James	$2000	$450000	15 years	6.2%
5	Kirra	$2650	$30000	30 years	7.9%

6 Juan and Denise borrow $450000 at a fixed interest rate of 8.34% per annum, compounded monthly. They plan to pay off the loan in exactly 25 years by making the same monthly repayments for 25 years.

How much will each monthly repayment be (nearest cent) and how much interest will they pay in total (nearest $10)?

7 Chris and Terri borrow $380000 at a fixed interest rate of 7.2% per annum, compounded monthly.

They plan to repay the loan by making the same regular repayments each month for twenty years.

How much will each repayment be (nearest cent) and how much will they repay in total (i.e. interest plus loan repayment) over the twenty years, to the nearest $10?

How much quicker would they have paid off the loan if they had paid $100 per month more?

INVESTIGATE

In the above cases we have only considered the cost of the house. However other expenses are incurred when purchasing a house, for example, stamp duty, insurance, settlement agent fees, purchaser's share of the rates. Investigate and write a brief report.

Miscellaneous exercise three

This miscellaneous exercise may include questions involving the work of this chapter, the work of any previous chapters, and the ideas mentioned in the Preliminary work section at the beginning of the book.

1 $8000 is invested for 5 years with an annual compound interest rate of 8%. Find the amount this account is worth at the end of the 5 years if the compounding occurs

   a   annually,   b   quarterly,   c   monthly.

2 What does it mean if the seasonal index for Autumn is 1.08 (or 108%)?

3 Deseasonalise the following raw data given the seasonal indices are as stated.
Give the deseasonalised data to the nearest integer.

Season	Mon	Tue	Wed	Thur	Fri	Sat
Raw data	113	79	84	151	151	160
Seasonal index	0.92	0.70	0.68	1.21	1.29	1.20

4 Find the first five terms of each of the following recursively defined sequences.

a $T_1 = 7$, $T_{n+1} = T_n + 12$.

b $T_1 = 100$, $T_{n+1} = T_n - 15$.

c $T_1 = 5000$, $T_{n+1} = 1.2 \times T_n$.

d $T_1 = 2000$, $T_n = T_{n-1} \div 10$.

e $T_1 = 4$, $T_n = 2T_{n-1} + 3$.

f $T_1 = 200$, $T_{n+1} = 1.5T_n - 4$.

5 Which of the following graphs show

a an outlier?   **b** seasonal variation?   **c** an increasing trend?

d a decreasing trend?   **e** no irregular fluctuations?

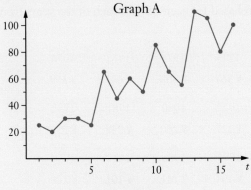

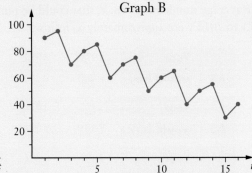

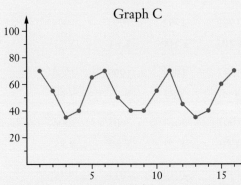

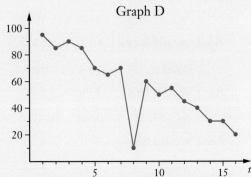

6 Use recurrence relation techniques for this question.

Johan borrows $50 000 with interest charged at 7.5% per annum on the declining balance, compounded monthly. At the end of the first period of one month, and at the end of every period of one month thereafter, Johan repays $1000 on the loan.

How much will Johan still owe on this loan immediately after the interest has been added and the $1000 has been credited for the end of

a month two?   **b** month six?   **c** year one?   **d** year two?

7 With the aid of a calculator with a built-in facility to carry out compound interest calculations, or otherwise, determine each of the following.

a The value at the end of five years of an initial investment of $12 000 invested at 6.4% per annum compound interest, compounded quarterly.

b The number of years required for an initial investment of $25 000 invested at 7.5% per annum compound interest, compounded annually, to be worth $75 000.

c The number of years required to pay off a loan of $400 000 if interest of 8% per annum is compounded monthly and $3500 is paid off the loan at the end of every month.

d The annual (nominal) compound interest rate required to see an initial investment of $25 000 exceed $125 000 within fifteen years if interest is compounded

    i annually,

    ii monthly.

8 In 1971 one Australian dollar, A$, could purchase approximately 400 Japanese yen.

In 2012 one Australian dollar could purchase approximately 80 Japanese yen.

The average numbers of yen, ¥, that could be purchased for each A$ 1 in each of the years from 1971 to 2012 were approximately as follows:

Year	1971	1972	1973	1974	1975	1976
A$ 1 would buy	¥ 400	¥ 372	¥ 367	¥ 408	¥ 405	¥ 379

Year	1977	1978	1979	1980	1981	1982
A$ 1 would buy	¥ 331	¥ 274	¥ 227	¥ 260	¥ 248	¥ 257

Year	1983	1984	1985	1986	1987	1988
A$ 1 would buy	¥ 233	¥ 211	¥ 193	¥ 141	¥ 101	¥ 100

Year	1989	1990	1991	1992	1993	1994
A$ 1 would buy	¥ 107	¥ 112	¥ 107	¥ 100	¥ 83	¥ 73

Year	1995	1996	1997	1998	1999	2000
A$ 1 would buy	¥ 70	¥ 77	¥ 90	¥ 86	¥ 78	¥ 68

Year	2001	2002	2003	2004	2005	2006
A$ 1 would buy	¥ 61	¥ 66	¥ 70	¥ 79	¥ 80	¥ 85

Year	2007	2008	2009	2010	2011	2012
A$ 1 would buy	¥ 93	¥ 99	¥ 75	¥ 81	¥ 82	¥ 81

(ABARES 2013, Agricultural commodity statistics 2013. CC BY 3.0)

With the assistance of a computer spreadsheet, display the above data graphically, together with the five-point moving averages.

Write a few sentences summarising the data.

How does today's exchange rate compare?

ISBN 9780170395069

4.

Finance II: Drawing down the investment

- Superannuation
- Annuities
- Indexing
- Perpetuities
- Suppose the frequency of payments ≠ the frequency of compounding
- Miscellaneous exercise four

Superannuation

Retirees in Australia may, if they satisfy certain conditions, qualify for the government age pension. The amount they receive depends on what other income and assets they have. However, the total of these pension payments form a considerable expenditure for the government and, with increasing life expectancy, an expenditure that could get bigger and bigger in years to come. For this reason the government encourages people to put money into savings during their years of employment so that they will have their own money to support themselves in retirement, and hence only need a part pension or perhaps even no pension payment at all from the government. Saving to have money available in retirement is called **superannuation**. During a person's working years they **accumulate funds** in their superannuation account and then, when they retire from work, they **draw down those funds** as an income they can live on, supplemented by the government pension if the requirements of an assets and income test are met.

INVESTIGATE

How much does a person need to have saved in superannuation to live comfortably in retirement?

What sort of annual income will a retired couple require?

What does the internet suggest as answers to these questions?

When a person is employed their employer must pay a certain percentage of their wage into a superannuation account. Thus during the employment years a person's superannuation account is in its *accumulation phase*. Then, when the employee retires, they draw money from the superannuation account as a self-funded pension, possibly supplemented by the government pension, to give them an income in retirement. The superannuation account is then in its *pension phase*. Of course in both the accumulation phase and the pension phase any funds in the account can earn interest. Thus in the pension phase interest is earned by the account and, at the same time, money is periodically withdrawn from the account.

EXAMPLE 1

At retirement Susie invests the $750 000 she has saved in her superannuation account in an account giving interest of 6% per annum compounded annually from which she will withdraw $60 000 after one year, and after each year thereafter. How much will be left in the account immediately after the tenth withdrawal (rounded to the nearest dollar)?

If she wants the account to last her at least 35 years, withdrawing a constant amount each year, what should this constant amount be (rounded down to a multiple of $100)?

Solution

Initial amount invested: $750 000

Account balance after 1 year: $750 000 × 1.06 − $60 000

Account balance after 2 years: ($750 000 × 1.06 − $60 000) × 1.06 − $60 000

We have
$$T_{n+1} = T_n \times 1.06 - \$60\,000,$$
and
$$T_0 = \$750\,000$$

Using the recursion capability of some calculators, a spreadsheet or the inbuilt financial programs of some calculators, we find that:

Immediately after the tenth withdrawal there will be $552 288 in the account, to the nearest dollar.

$$
\begin{array}{|l|}
\hline
\checkmark \quad a_{n+1} = 1.06 \cdot a_n - 60000 \\
\qquad\quad a_0 = 750000 \\[4pt]
\square \quad b_{n+1} = \square \\
\qquad\quad b_0 = 0 \\[4pt]
\square \quad c_{n+1} = \square \\
\qquad\quad c_0 = 0 \\
\hline
\end{array}
$$

n	a_n
6	6.5E+5
7	6.2E+5
8	6.0E+5
9	5.8E+5
10	5.5E+5

552288.075864291

Compound Interest

N	35
I(%)	6
PV	−750000
Pmt	51730.39423
FV	0
PpY	1
CpY	1

Varying the withdrawal amount to reduce the balance to zero in 35 years, or using a calculator with an inbuilt financial capability, we can determine that Susie can withdraw up to $51 700 each year (rounded down to a multiple of $100) for the account to last at least 35 years.

The graph below shows the declining balance in the account of the previous example which reduced the balance to zero in 35 years by withdrawing $51 730.39 per year.

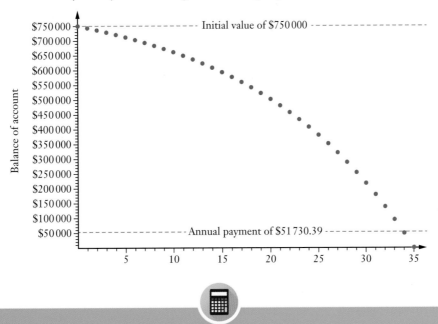

Can your calculator display a graph of this type from the terms of a recursively defined sequence? Investigate.

Annuities

In the previous example Susie managed her own *income stream* by making a deposit into an account, withdrawing a fixed amount each year to live on and allowing the balance of the account to earn interest. A compound interest investment from which regular payments are made for a fixed period of time is called an **annuity**.

Making regular payments to pay off a loan, as encountered in the previous chapter, is a bit like an annuity. The finance company invests money with us in the form of a loan and we make regular payments back. However the word annuity is usually used in the context of using a sum of money to set up a regular income.

By making a one-off payment a person can enter into an annuity arrangement whereby they will receive a guaranteed fixed regular amount for an agreed period of time. Such arrangements can be very useful for generating a steady income in retirement.

Fixed term annuities make regular payments for a fixed period of time. If the person who purchased the annuity dies before the end of the fixed period of time the remaining payments, or the remaining balance, would pass to a beneficiary.

Whole life annuities make regular payments to the purchaser of the annuity for as long as they live. Under this sort of annuity the payments stop when the person dies, i.e. upon the death of the purchaser such an annuity has no residual value. If the person dies soon after the annuity commences the finance company 'wins' but if the person lives well beyond normal life expectancy the person 'wins'. Life annuities can be seen as an insurance in case one lives a long time. Even if the person does live a long time, regular income is assured. The finance company spreads the risk of paying out for more years than expected by having annuities with a lot of people.

Indexing

In some cases the regular amount paid out from an annuity might increase over time so that the amount received keeps up with inflation and maintains its 'buying power'. We say that the regular payments are *indexed*, i.e. the payments change in line with the movement of some price index, e.g. the consumer price index, a measure of the increase in the price of a typical household selection of goods.

EXAMPLE 2

A person wishes to deposit $200 000 into an account that will earn interest of 8% per annum, compounded annually, withdrawing $25 000 at the end of the first year, $25 500 at the end of the second year, $26 010 at the end of the third year, and so on with each annual withdrawal being a 2% increase on the withdrawal of the previous year.

a What will be the balance of the account immediately after the fifth withdrawal?

b How long will it take for the balance in the account to reduce to zero?

Solution

a Initial amount invested: $200 000

Account balance after 1 year: $200 000 \times 1.08 - $25 000   I.e. $191 000

Account balance after 2 years: $191 000 \times 1.08 - $25 000 \times 1.02   I.e. $180 780

Account balance after 3 years: $180 780 \times 1.08 - $25 000 \times 1.02^2   I.e. $169 232.40

We have $T_{n+1} = T_n \times 1.08 - \$25\,000 \times 1.02^n$,

and $T_0 = \$200\,000$

Using a calculator able to display the terms of a recursively defined sequence, or a spreadsheet, or by working the terms through:

T_4,   the account balance after 4 years   = $156 240.79   (nearest cent)

T_5,   the account balance after 5 years   = $141 679.25   (nearest cent)

Immediately after the fifth withdrawal the balance will be $141 679.25.

b Continuing the sequence:

T_6,   the account balance after 6 years   = $125 411.57   (nearest cent)

T_7,   the account balance after 7 years   = $107 290.44   (nearest cent)

T_8,   the account balance after 8 years   = $87 156.53   (nearest cent)

T_9,   the account balance after 9 years   = $64 837.57   (nearest cent)

T_{10}, the account balance after 10 years = $40 147.26   (nearest cent)

T_{11}, the account balance after 11 years = $12 884.18   (nearest cent)

After a further year this balance will grow to $12 884.18 \times 1.08

                             = $13 914.91

The balance will reduce to zero at the end of the 12th year when the withdrawal will be just $13 914.91.

Note: Care needs to be taken when using online annuity calculators. Some annuities work on the basis that the first payout from the annuity occurs when the annuity is opened, e.g. as soon as a superannuation amount is rolled into pension phase, and some work on the basis that the first payout occurs at the end of the first month or year, whatever the agreed frequency of payment is.

Perpetuities

Suppose we invest $5000 in an account paying 8% per annum compounded annually, and we withdraw $400 at the end of each year.

Initial amount invested:	$5000	
Account balance after 1 year:	$5000 × 1.08 – $400	I.e. $5000.
Account balance after 2 years:	$5000 × 1.08 – $400	I.e. $5000.
Account balance after 3 years:	$5000 × 1.08 – $400	I.e. $5000. Etc.

In this case the annual withdrawal of $400 could continue forever because each year the interest earned exactly equals the amount paid out. This sort of annuity is called a **perpetuity**, the word perpetuity meaning *lasting forever*.

Consider the case, for example, of someone wishing to fund an annual award of $400 to be given to a deserving student in a school for some reason. By donating $5000 into an account paying 8% per annum the interest produced will fund the award forever, i.e. in perpetuity, endlessly.

Exercise 4A

1 An ex-pupil of a school wishes to invest a sum of money into an account paying 7.5% interest, compounded annually, so that every year thereafter $600 can be made available from the account as an award to a deserving student.

Explaining your reasoning determine how much money needs to be invested to make this perpetual award possible.

2 Someone who has made their fortune from mining wishes to set up an account that will allow a research grant of $75 000 to be paid every year thereafter to fund research at a University School of Mines Department. So that the grant can be perpetual the account, which pays 8% per annum interest, compounded quarterly, needs to generate annual interest of at least $75 000.

Explaining your reasoning determine how much the initial 'one-off' investment needs to be (rounded up to the next dollar).

3 What initial investment is required into an account that pays annual interest of 6%, compounded monthly, if the interest earned each year is to pay for an annual perpetual award of $15 000?

4 Let us suppose that $P is invested in an account that pays R% per annum, compounded annually. At the end of the first year, and every year thereafter, $A is withdrawn from the account and after exactly Y years the balance of the account reaches zero.

For each of the following, state the option that will increase Y, other quantities remaining the same.

a Increase P or decrease P.

b An increase in R or a decrease in R.

c An increase in A or a decrease in A.

5 At retirement, Kelvin invests the $620 000 he has in his superannuation account in an account giving interest of 5.8% per annum compounded annually, from which he will withdraw $50 000 at the end of every twelve months to live on. Using a recursive rule determine how much will be left in the account immediately after the tenth withdrawal (rounded to the nearest ten dollars).

For how many years will Kelvin be able to withdraw $50 000 per year?

For how many years would Kelvin have been able to withdraw $50 000 per year if instead the interest rate had been 7.8%?

6 Julie is left $765 000 in a will and decides to invest $700 000 of this money in an account paying 5.4% interest compounded annually. Julie plans to retire early and use this money to live on by withdrawing $50 000 at the end of the first year and each year thereafter until the account balance is exhausted. Use a recursive rule to determine for how many years will she be able to withdraw $50 000 from this account.

If instead Julie withdrew $45 000 at the end of each year for how many years could she do this?

Can you foresee any problems with the plan to withdraw a constant amount each year for a long period of time?

7 Use the financial capability of some calculators or computer programs to determine, to the nearest dollar, how much should be invested in an annuity earning 12% annual interest, compounded annually, to provide a regular annual income of $45 000 for exactly 25 years. (The first payment is to be at the end of the first 12 months).

8 Repeat the previous question but now for an interest rate of just 6% per annum compounded annually.

9 Use the financial capability of some calculators or computer programs to determine, to the nearest dollar, how much should be invested in an annuity earning 8% annual interest, compounded monthly, to provide a regular monthly income of $3000 for exactly 20 years. (The first payment is to be at the end of the first month).

10 Repeat the previous question but now for an interest rate of just 5% per annum compounded monthly.

ISBN 9780170395069

11 A finance company offers an annuity in which a person deposits a lump sum of $300 000 into an account and interest is compounded annually at the rate of 7.5% per year. At the end of the first year, and every year thereafter for as long as the balance in the account allows, the annuity pays out $30 000 with the final payment, which might be less than $30 000, reducing the balance to zero. (In the event of death the remaining funds will be paid into the estate of the deceased.)

   a Use a recurrence rule to determine how many $30 000 payments will be made and what the final payment will be (assuming the person does not die in the meantime).

   b Repeat part **a** but now determine the answer using a calculator with financial capability.

12 A finance company offers an annuity in which a person deposits a lump sum of $500 000 into an account and interest is compounded monthly at the rate of 0.4% per month. At the end of the first month and every month thereafter, the account pays out $4000 with the final payment, which might be less than $4000, reducing the balance to zero. (In the event of the death of the annuity holder remaining funds will be paid into the estate of the deceased.)

   a Use a recurrence rule to determine how many $4000 payments will be made and what the final payment will be (assuming the person does not die in the meantime).

   b Repeat part **a** but now determine the answer using a calculator with financial capability or computer based financial software.

13 To the nearest dollar, how much should be invested in an annuity paying 10% annual interest, compounded annually, to provide a regular annual income of $42 000 paid at the end of each year, for exactly 25 years?

   Determine the answer by

   a using a recursive formula with an initial 'guess' of $400 000

   b using the financial capability of some calculators or computer programs.

14 Tony wins $500 000 on the lottery and decides to use $400 000 of it to purchase a 15-year annuity that pays a fixed amount every year for fifteen years, commencing one year after the initial purchase. The annuity adds interest to the account balance at 8% per annum, compounded annually, and deducts the regular payment at the end of each year. At the end of the 15 years the account balance is zero.

   How much does Tony receive each year (to the nearest dollar)?

   Determine the answer by

   a using a recursive formula with an initial 'guess' of $50 000

   b using the financial capability of some calculators or computer programs.

15 A person wishes to deposit $300\,000 into an account that will earn interest of 8% per annum, compounded annually, withdrawing $30\,000 at the end of the first year, $30\,900 at the end of the second year, $31\,827 at the end of the third year, and so on with each annual withdrawal being a 3% increase on the withdrawal of the previous year.

a What will be the balance of the account immediately after the third withdrawal?

b How long will it take for the balance in the account to reduce to zero?

16 A person wishes to deposit $250\,000 into an account that will earn interest of 7.5% per annum, compounded annually, withdrawing $25\,000 at the end of the first year, $26\,000 at the end of the second year, $27\,040 at the end of the third year, and so on with each annual withdrawal being a 4% increase on the withdrawal of the previous year.

a What will be the balance of the account immediately after the eighth withdrawal?

b How long will it take for the balance in the account to reduce to zero?

Suppose the frequency of payments ≠ the frequency of compounding

So far in this chapter, in all of the annuity situations (apart from two questions involving perpetuities) the frequency of the payments, e.g. annual, quarterly, monthly, has been the same as the frequency of the compounding periods, e.g. annual, quarterly, monthly. Let us now consider situations in which compounding occurs more frequently than the payments are made (see example 3) and situations in which payments occur more frequently than compounding (see example 4)

EXAMPLE 3

Suppose that $250\,000 is invested into an account paying 8% interest, compounded quarterly, with $40\,000 withdrawn from the account at the end of the first year and at the end of every year thereafter.

Determine the balance of the account immediately after the sixth withdrawal.

Solution

Initial amount invested:	$250\,000	
Account balance after 1 year:	$250\,000 \times 1.02^4 - $40\,000	I.e. $230\,608.04.
Account balance after 2 years:	$230\,608.04 \times 1.02^4 - $40\,000	I.e. $209\,617.56 (nearest cent)

We have $$T_{n+1} = T_n \times 1.024 - \$40\,000,$$
and $$T_0 = \$250\,000$$

Displaying the terms of this sequence on a calculator, as shown on the right, we see that the balance of the account immediately after the sixth withdrawal is $106 866.65 (to the nearest cent).

Alternatively the same answer could be obtained using the inbuilt financial capability of some calculators: see the typical display below.

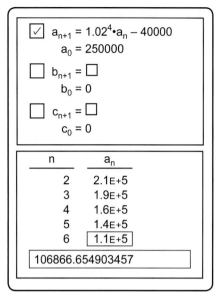

	$a_{n+1} = 1.02^4 \cdot a_n - 40000$
✓	$a_0 = 250000$
☐	$b_{n+1} = \square$
	$b_0 = 0$
☐	$c_{n+1} = \square$
	$c_0 = 0$

n	a_n
2	2.1E+5
3	1.9E+5
4	1.6E+5
5	1.4E+5
6	1.1E+5

106866.654903457

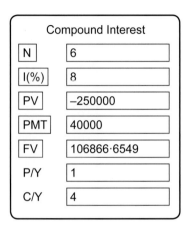

Compound Interest

N	6
I(%)	8
PV	−250000
PMT	40000
FV	106866·6549
P/Y	1
C/Y	4

Notice that in the previous example our use of the multiplication factor 1.02^4 brought our quarterly compounding up to its equivalent effective annual compounding and thereby had the frequency of compounding matching the frequency of payment.

Now suppose that payments occur more frequently than compounding. For example payments occurring every quarter, compounding occurring annually and an annual interest rate of 8%. We can again adjust the compounding period to match the payment period by finding the quarterly multiplication factor that will produce an annual compounding of 8%. This quarterly multiplication factor is 1.019 426 547 because,

$$\text{if } x^4 = 1.08 \qquad \text{then} \qquad x = 1.019\,426\,547.$$

However, whilst we would need this equivalent multiplication factor if we were using a recursive approach, the use of a calculator with built-in financial capability gives a ready alternative that avoids us having to calculate this figure, as shown in the next example.

Suppose that $250\,000 is invested into an account paying 8% interest, compounded annually, with $10\,000 withdrawn from the account three months later and a further $10\,000 each quarter thereafter.

Determine the balance of the account after three years.

Solution

The display shows that after 3 years (12 quarterly payments of $10\,000, i.e. 4 payments per year) an account paying 8% interest per annum, compounded annually (1 compounding per year), and initially worth $250\,000, will be worth $181\,238.775 after three years.

Compound Interest	
N	12
I(%)	8
PV	−250000
PMT	10000
FV	181238·775
P/Y	4
C/Y	1

Notice that using the 1.019426547 multiplication factor mentioned on the previous page to create an appropriate recursive rule gives this same answer, as shown below. However, for these questions, in which the payments occur more frequently than the compounding, using the financial capability of some calculators is recommended rather than using a recurrence relation.

$a_{n+1} = 1.019426547 \cdot a_n - 10000$
$a_0 = 250000$

n	a_n
8	2.1E+5
9	2.0E+5
10	1.9E+5
11	1.9E+5
12	1.8E+5

181238.775301663

Shutterstock.com/simez78

RETIREMENT

Exercise 4B

1 Use a recursive approach for this question.

Suppose that $400\,000 is invested into an account paying 6% per annum interest, compounded semi-annually, with $35\,000 withdrawn from the account at the end of the first year and at the end of every year thereafter.

Determine the balance of the account immediately after the tenth withdrawal.

2 For this question use a recursive approach (or a spreadsheet of your own creation) and make $500\,000 your 'first guess' at the amount invested.

To the nearest dollar, how much needs to be invested in an account paying interest of 6% per annum, compounded quarterly, for $45\,000 to be withdrawn at the end of every year for 25 years?

3 For this question use the financial capability of some calculators.

How much needs to be invested, to the nearest cent, in an account paying interest of 9% per annum, compounded annually, for $10\,000 to be paid at the end of the first 3 months, and every three months thereafter for 15 years, at the end of which time the account has zero balance remaining?

4 For this question use the financial capability of some calculators.

How much needs to be invested, to the nearest cent, in an account paying interest of 6% per annum, compounded annually, for $4000 to be paid at the end of the first month, and every month thereafter for 20 years, at the end of which time the account has zero balance remaining?

5 Upon retirement John switches his superannuation fund of $480\,000 into an account paying 9% per annum compounded monthly. After three months, and every three months thereafter John wants the account to pay him $15\,000, for as long as possible.

How many payments of $15\,000 will this scheme allow John to receive and what will be the final payment that closes the account?

RESEARCH

Do some research and write a brief report on some of the following questions.

• What incentives and rules does the Government of Australia put in place to ensure / encourage people to put money into superannuation?

• What is the current average balance that people have in superannuation?

• Why is an ageing population a challenge for governments?

• What is the retirement age in Australia? What is it in other countries?

Computer spreadsheet or graphic calculator activity

Sue wins \$120 000 in a lottery. Rather than spend it all immediately she decides to use some of it, and perhaps even all of it, to allow herself to take some years off work, or at least reduce her hours somewhat.

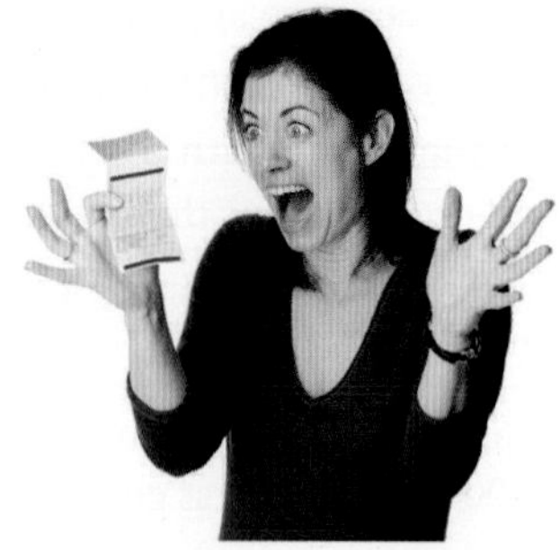

She asks for your advice regarding how long the money would last under each of five schemes she has in mind.

Determine how many months she could withdraw the stated monthly withdrawal in each of the following schemes.

Scheme 1 Put the money in a box under the bed and withdraw \$1500 each month to live on.

Scheme 2 Invest the entire \$120 000 in an account paying 8% interest per year, compounded monthly, and withdraw \$1500 at the end of each month to live on.

Scheme 3 Same as scheme 2 but this time suppose the account pays interest of 9% per annum compounded monthly.

Scheme 4 Use \$20 000 of the \$120 000 to pay for flights and accommodation for an around the world holiday, invest the rest in an account earning 9% per annum interest, compounded monthly, and withdraw \$1500 each month during the holiday and thereafter.

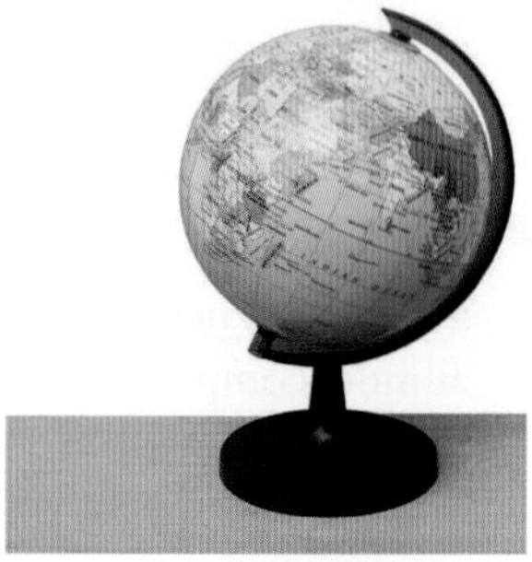

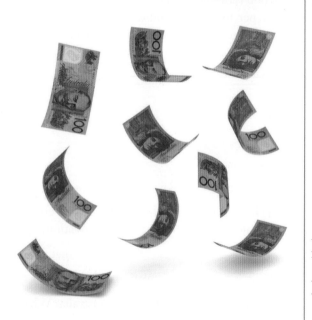

Scheme 5 Invest the entire \$120 000 in an account paying 9% interest per year, compounded monthly. At the end of each of the first 12 months withdraw \$1500, at the end of each of the next 12 months withdraw \$1600, at the end of each of the next 12 months withdraw \$1700 and so on, increasing the monthly withdrawal by \$100 after each 12-month period.

ISBN 9780170395069

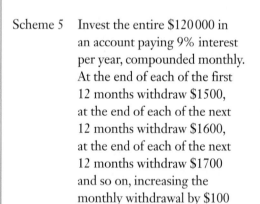

Miscellaneous exercise four

This miscellaneous exercise may include questions involving the work of this chapter, the work of any previous chapters, and the ideas mentioned in the Preliminary work section at the beginning of the book.

1 How much interest is earned in four years if $4000 is invested in an account paying 12% per annum compounded monthly?

2 A time series graph showing the number of people attending a particular music festival each year, for ten years, suggested that linear regression could be used to summarise the relationship between N, the numbers attending, and t the year number where $t = 1$ is used for the first year in the ten-year period.

The least squares regression line for the data had equation:

$$N = 855t + 9459$$

Interpret the '855' in this equation.

3 Giving each answer as a percentage correct to three decimal places, calculate the effective interest rate if the annual nominal rate is 6% and interest is compounded

a   quarterly,   **b**   monthly,   **c**   daily.

4 To the nearest cent, what 'one-off' lump-sum payment must be invested into an account paying compound interest of 8% per annum compounded quarterly for the investment to be worth $5000 two years later?

5 If we use the recursion relation $T_{n+1} = T_n \times 1.065$, $T_0 = \$4300$, the value of T_n gives the value of an account n years after a deposit was made to open the account, with interest compounded annually.

a   What was the amount of the deposit?

b   What was the annual interest rate?

c   Find the balance of the account after interest had been added at the end of the sixth year (to the nearest cent)?

d   What annual interest rate would achieve the balance that was the answer to part **c** but after just two years? (Give your answer as a percentage correct to three decimal places).

6 The table below relates to a savings account which Waylon opens by making an initial deposit and then adding a further fixed amount at the end of each month.

Interest earned is compounded monthly with the same monthly interest rate throughout.

Month	Amount in account at beginning of month	Amount in account at end of month, following adding of interest and deposit
1	$1600	$1874.00
2	$1874.00	$2152.11

Determine the initial investment, the fixed monthly interest rate and the fixed monthly deposit.

7 a Determine the values that go in the spaces marked A, B, … G in this table.

t	Year	Months of year	Sales	Four month mean for year	Sales ÷ (four month mean for the year)
1		First 4 months	442		1.1190
2	2010	Second 4 months	299	395	A
3		Third 4 months	444		1.1241
4		First 4 months	307		1.0233
5	2011	Second 4 months	241	300	B
6		Third 4 months	352		1.1733
7		First 4 months	294		1.1264
8	2012	Second 4 months	203	261	C
9		Third 4 months	286		1.0958
10		First 4 months	213		0.9953
11	2013	Second 4 months	171	214	D
12		Third 4 months	258		1.2056
13		First 4 months	170		1.0625
14	2014	Second 4 months	122	E	F
15		Third 4 months	188		G

b Create a completed version of the following tables.

Calculation of seasonal indices

	2010	2011	2012	2013	2014	Seasonal index (3 decimal places)
1st 4 months	1.1190	1.0233	1.1264	0.9953	1.0625	1.065
2nd 4 months						
3rd 4 months	1.1241	1.1733	1.0958	1.2056		

Seasonally adjusted sales figures, S (rounded to the nearest integer)

	2010	2011	2012	2013	2014
1st 4 months					
2nd 4 months					
3rd 4 months					

c Use the ordered pairs (t, S) to determine the equation of the regression line
$$S = At + B.$$

d Use this regression equation and the seasonal indices to predict sales for the three 4-month intervals of 2015.

5.

Minimum spanning trees

Spanning trees

The network on the right shows the roads that exist between nine farms, A to I, with the numbers indicating distances in kilometres. The water authority wish to connect these farms to the mains water supply by laying water pipes alongside existing roads.

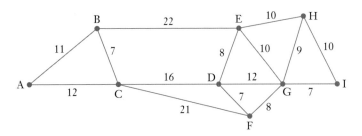

This is not the same as finding the shortest route from A to I because this shortest route, shown on the right, does not connect up all the farms (B, E, F and H are not 'on line').

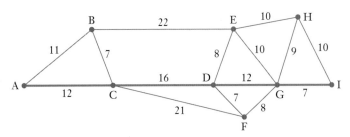

Instead we need to find a system that includes all the farms but which has no unnecessary connections. What we need is a **minimum spanning tree** for this network.

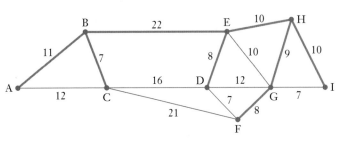

(A spanning tree of a graph is a subgraph that connects all of the vertices and is itself a tree.)

The spanning tree shown requires a total length of 85 km (= 11 km + 7 km + 22 km + 8 km + 10 km + 9 km + 10 km + 8 km).

Can you find a shorter spanning tree?
Can you find the shortest spanning tree?
How long is it?

If you think you have found the minimum spanning tree for the network shown above, then try to find it for the network shown below.

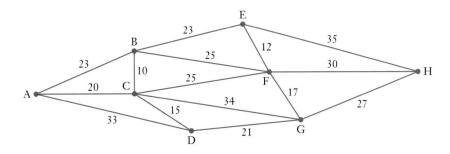

ISBN 9780170395069

Minimum spanning trees – a systematic approach

Consider the network on the right.

To determine the minimum spanning tree systematically we can in fact start at any vertex but we will start at A as that seems a logical place to start.

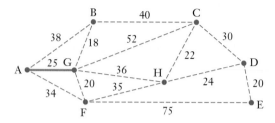

From A we next bring 'on line' the vertex that is A's *nearest neighbour*. Point G, being just 25 units from A is the nearest neighbour in this example.

Now that we have A and G on line we next connect to the vertex that is closest to one of these two vertices. In our example, B is the next nearest neighbour to one of our on line points A and G.

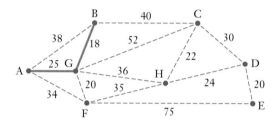

Now that we have A, G and B on line we now look for the vertex that is not yet on line but that is the nearest neighbour to one of our on line points.

Thus point F, being just 20 units from G, is the next nearest neighbour.

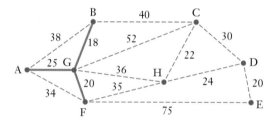

We now have A, G, B and F on line and again look for the next nearest neighbour. In this case point A is only 34 units from F but we already have A on line so this connection would be pointless. (It would give an unwanted *cycle* in our connection.) Thus we choose to bring H on line as it is the nearest neighbour not yet on line.

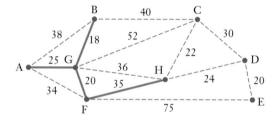

Continuing in this way until all points are on line gives the minimum spanning tree shown on the right.

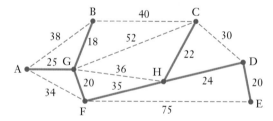

Of course this process would usually be carried out on one copy of the network. The five copies used here are purely to illustrate the steps of the method.

The reader should now use this method on a copy of the network shown on the right but this time start at a vertex other than A and confirm that a different starting point still leads to the same minimum spanning tree.

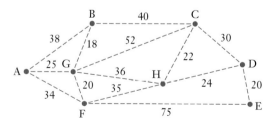

The method explained on the previous page, in which we:

choose any vertex as our initial 'on line' vertex and then build up the spanning tree by connecting online vertices to the 'nearest neighbour', whilst always making sure that no cycles are introduced,

is called **Prim's algorithm**.

An alternative approach is to use **Kruskal's algorithm**. In this method we:

start with the 'shortest' edge, then choose the next' shortest' edge, even if it is not connected to the previously selected edge, and then continue this process, always making sure that no cycles are introduced. When all vertices are 'on line' we have the minimum spanning tree.

Using this Kruskal's approach for the network on the previous page:

First we choose the 'shortest' edge in the network, in this case GB which has a weighting of just 18.

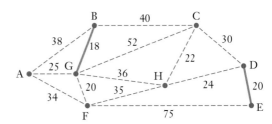

Then we choose the next 'shortest' edge. In this case there are two, GF and DE, each with a weighting of 20. It does not matter which of these two we choose. We will choose DE, as shown.

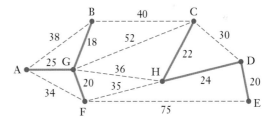

Continuing this process sees us choose GF (20) then HC (22) then HD (24) and then AG (25) as our next four edges, as shown on the right.

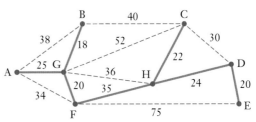

The next 'shortest' edge will be CD (30). However this will not bring any vertex on line that is not already on line. (It will give an unnecessary cycle.)

The next 'shortest' after CD is AF (34) which again introduces an unnecessary cycle.

Choosing the next 'shortest' which does not introduce a cycle we choose FH (35), and so the minimum spanning tree is completed.

Exercise 5A

Find the minimum spanning tree for each of numbers **1** to **6**, and state its length in each case.

1

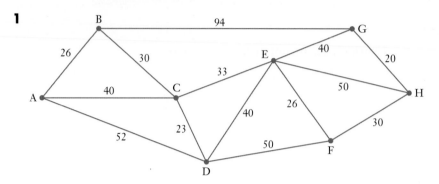

2

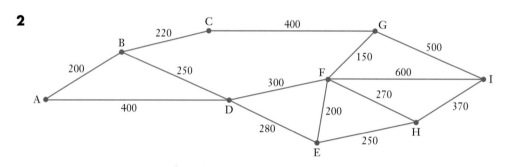

3

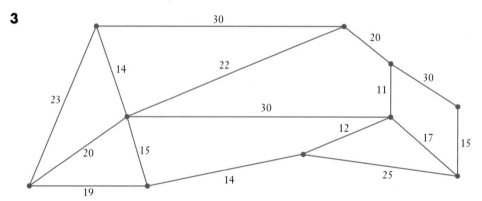

4

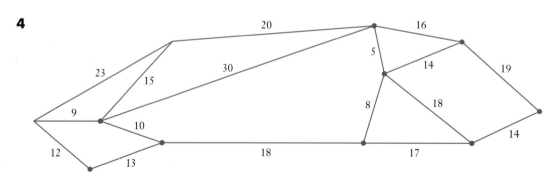

ISBN 9780170395069

5

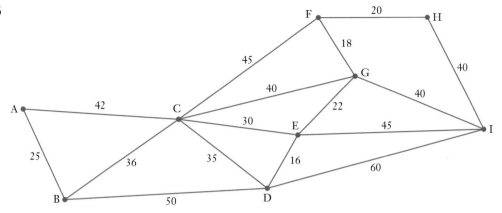

6

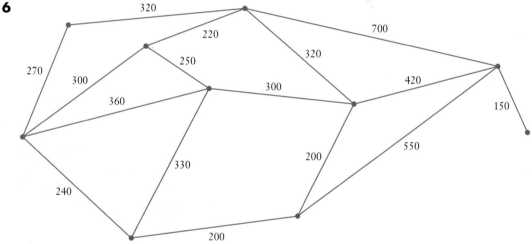

7 A market gardener wishes to replace the old piping currently connecting his seven-outlet irrigation system. Rather than replacing all of the existing piping he wishes to replace the system with the minimum total length of piping that will keep all seven outlets connected, either directly or indirectly. He will then close off those unnecessary parts of the existing system.

The graph on the right shows the existing system. Points A to G are the outlets and the numbers indicate the length of each section in metres.

Which sections should he replace with new piping, which should he close off, and what total length of new piping will be needed?

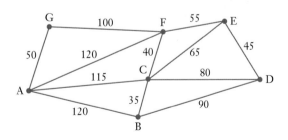

5. Minimum spanning trees ●●●●●●●●

8 A camp-site is being designed with three ablution blocks, A_1, A_2 and A_3, two play areas for children, P_1 and P_2, two barbecue areas, B_1 and B_2, and a shop, S. The sketch below shows the relative locations of these features with the distances between them given in metres.

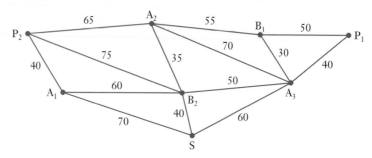

Paths are to be built so that all of these features are linked. On a copy of the above diagram indicate the network of paths that spans all of the features whilst minimising the total length of pathway.

What is this minimum length of pathway?

9 The network below shows the system of tunnels that exist linking the seven entrances of an old silver mine. The numbers on the edges show the length of each section of tunnel in metres.

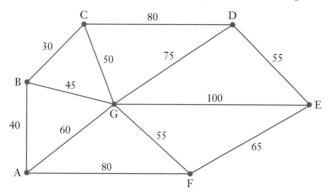

The mine is a popular tourist attraction but all of the tunnels are in need of repair if they are to remain open for visitors.

The owners cannot afford to repair all of the tunnels so instead decide to repair the minimum total length possible, while still allowing entrance to the repaired system to be possible from all seven of the existing entrances.

The tunnels not being repaired will then be closed off to visitors.

They ask you to determine which tunnels should be repaired and which shut down.

a On a copy of the network show the tunnels the owners should repair.

b What is the total length of tunneling that will be *closed off* to visitors?

ISBN 9780170395069

10 The table below shows the distances, in km, along the roads that exist between seven towns:
Alsop, Beal, Clan, Dawse, Erle, Felix, and Grant.

(A '–' indicates that there is no direct road between those towns.)

	Alsop	Beal	Clan	Dawse	Erle	Felix	Grant
Alsop	–	–	310	–	300	–	620
Beal	–	–	550	430	530	330	–
Clan	310	550	–	–	350	–	–
Dawse	–	430	–	–	–	210	390
Erle	300	530	350	–	–	380	360
Felix	–	330	–	210	380	–	–
Grant	620	–	–	390	360	–	–

The sketch on the right shows the approximate
positions of the seven towns with respect to
each other.

● Clan   ● Beal

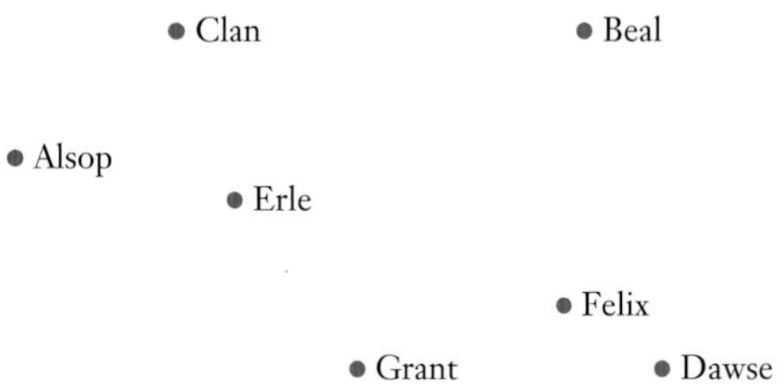

a Make a copy of this sketch and show on
your copy the roads that exist between
the towns and the lengths of these roads.

● Alsop

● Erle

b Water pipes are to be laid along some of
these roads so that the seven towns are, either
directly or indirectly, connected to each other.

● Felix

● Grant   ● Dawse

Determine the minimum length of piping required.

Minimum spanning tree from the distances table

In the last question of the previous exercise it was reasonably easy to transfer the details given in the
table onto a network diagram because we were given a diagram showing the approximate locations of
the towns with respect to each other. We then used the network diagram to determine the minimum
spanning tree. However, if the network involved had been larger, and/or the relative locations of the
vertices not have been known, the task of creating the network diagram with all of the weightings in
place would have been more difficult. Fortunately it is possible to determine the minimum spanning
tree directly from the table of distances without having to create the network diagram first. This
process is demonstrated on the next page. However, as you read through it you should notice that the
method is actually Prim's algorithm …

*choose any vertex as our initial 'on line' vertex and then build up the spanning tree by connecting on line
vertices to the 'nearest neighbour' whilst always making sure that no cycles are introduced*

applied to the table of distances rather than to the network drawing.

Consider the table of distances shown on the right, which is actually the one from the last question of the previous exercise.

	A	B	C	D	E	F	G
A	–	–	310	–	300	–	620
B	–	–	550	430	530	330	–
C	310	550	–	–	350	–	–
D	–	430	–	–	–	210	390
E	300	530	350	–	–	380	360
F	–	330	–	210	380	–	–
G	620	–	–	390	360	–	–

Whilst we can start at any vertex we will choose to start at A.

With A *on line* we look down the A column to see which is the shortest connection that can be made from A.

In this case it is the 300 km connection to town E.

↓

	A	B	C	D	E	F	G
A	–	–	310	–	300	–	620
B	–	–	550	430	530	330	–
C	310	550	–	–	350	–	–
D	–	430	–	–	–	210	390
E	(300)	530	350	–	–	380	360
F	–	330	–	210	380	–	–
G	620	–	–	390	360	–	–

Having chosen to look down the *A column* for a connection *from* A we then rule a line through the A *row* to ensure that we do not, at some later stage, use this line to make an unnecessary connection back *to* A.

(Alternatively, we could have chosen to look across row A for the shortest connection, and then rule a line through *column* A.)

↓

	A	B	C	D	E	F	G
~~A~~			~~310~~		~~300~~		~~620~~
B	–	–	550	430	530	330	–
C	310	550	–	–	350	–	–
D	–	430	–	–	–	210	390
E	(300)	530	350	–	–	380	360
F	–	330	–	210	380	–	–
G	620	–	–	390	360	–	–

Having selected the 300 km connection to vertex E we now have both A and E *on line* and indicate this by the arrows at the top of these two columns.

Again, to avoid making unnecessary connections at a later stage we rule a line through the E *row*.

We now look down our two *on line* columns, A and E, and choose the minimum connection that can be made from one of these two vertices.

↓   ↓

	A	B	C	D	E	F	G
~~A~~			~~310~~		~~300~~		~~620~~
B	–	–	550	430	530	330	–
C	310	550	–	–	350	–	–
D	–	430	–	–	–	210	390
~~E~~	(300)	~~530~~	~~350~~			~~380~~	~~360~~
F	–	330	–	210	380	–	–
G	620	–	–	390	360	–	–

ISBN 9780170395069

This gives us the 310 km road from A to C as our next connection.

This brings vertex C on line and we rule a line through the C row.

	A	B	C	D	E	F	G
A			~~310~~		~~300~~		~~620~~
B	–	–	550	430	530	330	–
C	(310)	~~550~~	~~–~~	~~–~~	350	~~–~~	~~–~~
D	–	430	–	–	–	210	390
E	(300)	~~530~~	~~350~~			~~380~~	~~360~~
F	–	330	–	210	380	–	–
G	620	–	–	390	360	–	–

We now look down our three *on line* columns, A, C and E, and choose the minimum connection that can be made from one of these three vertices.

This gives us the 360 km road from E to G as our next connection.

	A	B	C	D	E	F	G
A			~~310~~		~~300~~		~~620~~
B	–	–	550	430	530	330	–
C	(310)	~~550~~			350		
D	–	430	–	–	–	210	390
E	(300)	~~530~~	~~350~~			~~380~~	~~360~~
F	–	330	–	210	380	–	–
G	620	–	–	390	(360)	–	–

This brings town G on line and we rule a line through the G row.

We continue this process until all of the towns are on line. The circled items in the table indicate the direct connections that together form the minimum spanning tree.

	A	B	C	D	E	F	G
A			~~310~~		~~300~~		~~620~~
B	–	–	550	430	530	330	–
C	(310)	~~550~~			350		
D	–	430	–	–	–	210	390
E	(300)	~~530~~	~~350~~			~~380~~	~~360~~
F	–	330	–	210	380	–	–
G	~~620~~			~~390~~	(360)		

From the completed table on the right we see that the roads AC, AE, EF, EG, FB and FD form the minimum spanning tree, with a total length of 1890 km.

The reader should confirm that this is the same answer as was obtained for question 10 of Exercise 5A.

This application of **Prim's algorithm** may at first seem rather tedious and lengthy but you will find that it can be carried out quite quickly and, of course, in practice it only requires one copy of the table. The multiple copies shown here were to allow each step of the process to be clearly seen.

	A	B	C	D	E	F	G
A			~~310~~		~~300~~		~~620~~
B			~~550~~	~~430~~	~~530~~	(330)	
C	(310)	~~550~~			350		
D		~~430~~				(210)	~~390~~
E	(300)	~~530~~	~~350~~			~~380~~	~~360~~
F		~~330~~		~~210~~	(380)		
G	~~620~~			~~390~~	(360)		

Exercise 5B

Determine the length of the minimum spanning trees for the networks defined by the tables in questions **1** to **5**. (A '–' in the table indicates there is no direct route between the locations.)

1

	A	B	C	D
A	–	–	21	20
B	–	–	39	35
C	21	39	–	24
D	20	35	24	–

2

	A	B	C	D	E
A	–	38	–	62	32
B	38	–	37	–	35
C	–	37	–	40	22
D	62	–	40	–	41
E	32	35	22	41	–

3

	A	B	C	D	E	F
A	–	450	–	–	–	470
B	450	–	300	–	–	400
C	–	300	–	500	200	400
D	–	–	500	–	420	700
E	–	–	200	420	–	310
F	470	400	400	700	310	–

4

	A	B	C	D	E	F	G
A	–	2.6	–	–	–	3.6	4.0
B	2.6	–	5.0	–	–	–	2.8
C	–	5.0	–	5.0	3.0	–	3.5
D	–	–	5.0	–	5.3	–	–
E	–	3.0	5.3	–	5.7	3.9
F	3.6	–	–	–	5.7	–	2.8
G	4.0	2.8	3.5	–	3.9	2.8	–

5

	A	B	C	D	E	F	G	H	I
A	–	650	–	–	–	370	400	–	–
B	650	–	600	–	–	–	340	350	–
C	–	600	–	220	–	–	–	–	380
D	–	–	220	–	560	–	–	–	390
E	–	–	–	560	–	700	–	300	210
F	370	–	–	–	700	–	–	410	–
G	400	340	–	–	–	–	–	–	–
H	–	350	–	–	300	410	–	–	250
I	–	–	380	390	210	–	–	250	–

6 Determine the length of the minimum spanning tree for the network defined by the table on the right. (A '–' in the table indicates there is no direct route between the locations.)

Name the five roads that would be used in your minimum spanning tree.

	A	B	C	D	E	F
A	–	8.7	–	8.9	–	9.1
B	8.7	–	9.5	–	–	4.5
C	–	9.5	–	6.1	7.0	5.3
D	8.9	–	6.1	–	3.5	–
E	–	–	7.0	3.5	–	3.8
F	9.1	4.5	5.3	–	3.8	–

ISBN 9780170395069

7 Eight farmlets (A, B, C, D, E, F, G and H) are to be connected to mains water supply. The water authority plans to lay pipes alongside some of the existing roads that link the farmlets. The plan is to bring all eight farmlets 'on line' using those roads that form the minimum spanning tree for the network of existing roads. In this case 'minimum' meaning the minimum total length. The lengths of the existing roads are as in the table below. All lengths are given in kilometres and a '–' in a space means that there is no existing road directly linking those two places.

Determine the minimum length of pipe required.

	A	B	C	D	E	F	G	H
A	–	4.2	–	–	7.2	4.1	–	–
B	4.2	–	6.0	–	–	2.0	–	–
C	–	6.0	–	4.4	–	6.3	3.0	–
D	–	–	4.4	–	6.8	–	5.0	6.0
E	7.2	–	–	6.8	–	–	–	2.3
F	4.1	2.0	6.3	–	–	–	3.5	2.8
G	–	–	3.0	5.0	–	3.5	–	2.5
H	–	–	–	6.0	2.3	2.8	2.5	–

iStock.com/GOlFX

8 A school has six teaching areas, A, B, C, D, E and F, that need to be linked for the school video, public address, internet and phone systems.

Cables are to be installed to create a spanning tree for the network with vertices at A, B, C, D, E and F.

Find the minimum cost of the spanning tree if the cost of the direct cable links between vertices are as in the table on the right.

	A	B	C	D	E	F
A	0	$170	$525	$410	$210	$325
B	$170	0	$400	$310	$145	$225
C	$525	$400	0	$140	$315	$200
D	$410	$310	$140	0	$190	$100
E	$210	$145	$315	$190	0	$115
F	$325	$225	$200	$100	$115	0

iStock.com/izusek

9 A railway authority wishes to build a freight railway network linking seven towns. The network is to be constructed so that it will be possible to travel by rail from any one of the towns to any of the other six in the network, perhaps not directly, but at least by going via other towns in the network. The towns are Harlow, Gatley, Olber, Wayley, Raine, Lewis and Fitch and the distances involved for the feasible rail routes from any of these towns to any of the others are given in the following table (in kilometres).

	Harlow	Gatley	Olber	Wayley	Raine	Lewis	Fitch
Harlow	0	300	510	316	122	195	320
Gatley	300	0	224	100	212	378	206
Olber	510	224	0	200	400	517	269
Wayley	316	100	200	0	202	327	112
Raine	122	212	400	202	0	189	198
Lewis	195	378	517	327	189	0	252
Fitch	320	206	269	112	198	252	0

a Based on these distances determine the minimum length of track required for this network.

b In this 'minimum length network' which six pairs of towns would have a direct rail link? (By direct rail link we mean able to journey from one town to the other without having to pass through any of the other towns on the way.)

10 The table below indicates the distances, in kilometres, along roads that exist between ten locations, A to J. (A dark shaded cell in the table indicates that there is no direct road between the locations.)

	A	B	C	D	E	F	G	H	I	J
A				60	30		70			
B					45	20				40
C				65				28	35	30
D	60		65		46	18	20	43		
E	30	45		46		42				
F		20		18	42					41
G	70			20				38		
H			28	43			38		54	
I			35					54		
J		40	30			41				

The roads are in need of upgrading and it is decided that, rather than upgrade all of the roads, only those roads needed to ensure that all ten locations are linked by upgraded roads, either directly or indirectly, will be upgraded.

Which roads should be upgraded if the total length of road to be upgraded is to be kept to a minimum and what would this minimum length be?

ISBN 9780170395069

This miscellaneous exercise may include questions involving the work of this chapter, the work of any previous chapters, and the ideas mentioned in the Preliminary work section at the beginning of the book.

1 The following employment data gives the number of people in full-time employment in Australia each quarter from January 2010 to April 2013, with the exception of July 2011 for which the information is missing from the table.

Data point (n)	1	2	3	4	5	6	7
Month	Jan 2010	April 2010	July 2010	Oct 2010	Jan 2011	April 2011	July 2011
Employed ('000) (N)	7630	7677	7749	7821	7876	7897	

Data point (n)	8	9	10	11	12	13	14
Month	Oct 2011	Jan 2012	April 2012	July 2012	Oct 2012	Jan 2013	April 2013
Employed ('000) (N)	7911	7942	7971	7989	8024	8028	8022

(Based on Australian Bureau of Statistics data.)

a Use linear regression techniques to determine the equation of the least squares line of best fit, $N = an + b$.

b Interpret your value of a in your answer for part **a**.

c Use your regression line to predict the employment figure for July 2011 and for January 2015 stating which answer you would expect to be the more reliable and why.

d View the given data on a graphic calculator or computer and write some appropriate comments.

2 In an attempt to determine how long an investment of $350 000 placed in an account paying 6% per annum, compounded monthly, would last if $1500 was withdrawn after one month and at the end of every month thereafter, Jim put this information into a financial calculator, as shown below left. When he asked the calculator to determine N, the number of withdrawals the calculator returned 'Error', as shown below right. Explain.

Compound Interest	
N	
I%	6
PV	−350000
PMT	1500
FV	0
P/Y	12
C/Y	12

Compound Interest	
N	Error
I%	6
PV	−350000
PMT	1500
FV	0
P/Y	12
C/Y	12

3 The table below shows some of the data for the number of kilolitres of water used by a household each quarter for a period of three years.

t	Year	Quarter	No. of kL	Uncentred 4-pt MA	Centred 4-pt MA
1	One	1st	95		–
				–	
2		2nd	60		–
				a	
3		3rd	36		d
				b	
4		4th	77		78
				c	
5	Two	1st	131		e
				82	
6		2nd	f		81
				g	
7		3rd	44		79
				h	
8		4th	i		k
				j	
9	Three	1st	123		m
				l	
10		2nd	92		84
				n	
11		3rd	60		–
				–	
12		4th	o		–

Calculate the value of **a, b, c, d, … o**.

4 To the nearest cent, what 'one-off' lump-sum payment must be invested into an account paying compound interest of 8% per annum, compounded monthly, for the investment to be worth $5000 four years later?

5 Amy takes out a loan of $15 000. Interest of 8.2% of the outstanding balance is added at the end of each year and then Amy's regular annual repayment of $2500 is taken from the outstanding balance.

a How much does Amy still owe on this loan immediately after she makes the $2500 repayment at the end of year six?

b If instead Amy wishes to pay the loan off in the six years what does her regular annual repayment need to be?

ISBN 9780170395069

6.

Maximum flow

- Maximum flow – a systematic approach
- Maximum flow = minimum cut
- Miscellaneous exercise six

Situation

Suppose the network on the right shows the arrangement of water pipes around a farm with the numbers indicating the maximum rate of flow that each pipe can manage, in litres per minute.

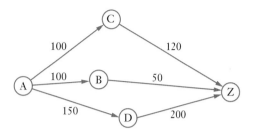

Point A is the source of the water supply. The three pipes coming from A could draw water from A at a total rate of 350 litres every minute but could this amount of water be delivered to Z each minute?

What is the maximum amount of water that can be delivered to Z each minute?

Now suppose we add a new pipe with a capacity of 50 litres per minute. The networks below show two ways that this could be done. What is the maximum flow from A, the **source**, to Z, the **sink**, in each of these situations?

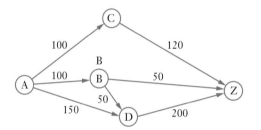

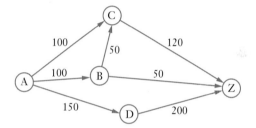

When the network is reasonably simple we can determine the maximum flow by mental reasoning as we have done above. However, for more complicated networks we need a procedure, or algorithm, to follow. Before such an algorithm is suggested for you, try to develop a procedure for yourself to determine the maximum flow through the network below from a source at A to a sink at G. The network shows the units of electrical power that can be delivered along the various links between relay stations.

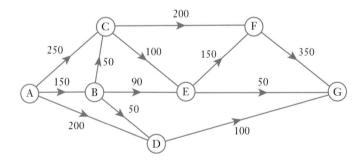

Note: In reality, in situations like these there is often more than one *source* from which we can obtain the water or the power or whatever it is that is involved. In addition there is frequently more than one *sink* where this water or power is required. However, in this book we will restrict our attention to situations in which there is just one source and just one sink.

Maximum flow – a systematic approach

Suppose we wish to determine the maximum flow that can be achieved for the network shown below from the source at A to the sink at H.

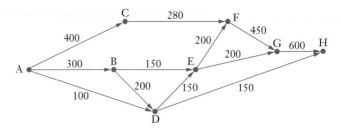

To approach the problem systematically we consider the separate routes in turn, from the top of the diagram to the bottom. We send the maximum flow we can along each route and note the remaining capacity of each pipe involved.

Consider route ACFGH.

This route can carry a maximum of 280 units (because that is all that CF can carry).

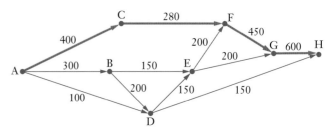

Sending 280 units along ACFGH leaves each pipe involved with the carrying capacity indicated in the next diagram. (Check carefully that you agree with these figures and can see how they are obtained.)

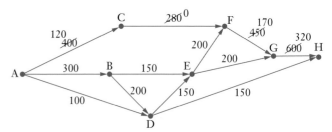

Now consider ABEFGH.

We can send 150 units along this route.

The remaining capacity of each pipe will then be as shown on the right.

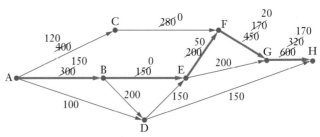

We can send 20 units along ABDEFGH.

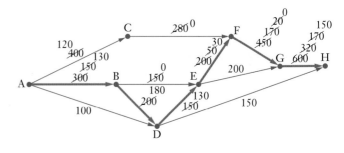

We can send 130 units along ABDEGH.

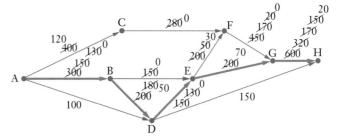

We can send 100 units along ADH.

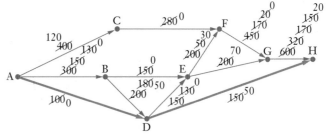

Thus the flow is:

ACFGH	280 units	+
ABEFGH	150 units	
ABDEFGH	20 units	
ABDEGH	130 units	
ADH	100 units	
giving a total of	680 units	

The maximum flow from the source at A to the sink at H is 680 units.

Note:
- This process would normally be carried out using a single diagram as shown in the example on the next page.

- The arcs leaving A have a total capacity of 800 units (= 400 + 300 + 100). Our maximum flow of 680 would indicate that there must be 120 units that we have not been able to use. This can be seen in the diagram above as 120 units of unused capacity in AC.

  Similarly the arcs going to the sink, H, have the capacity to deliver 750 units (= 600 + 150). Our maximum flow would indicate there must be 70 units that we have not been able to use. This can be seen in the diagram above as 50 units of unused capacity in DH and 20 units of unused capacity in GH.

- By noting the unused capacity in each arc we can determine the flow along each pipe to obtain maximum flow:

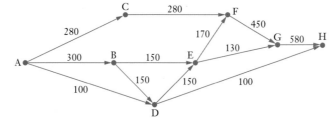

  Notice in this diagram that 680 units (the maximum flow) leaves A, 680 units arrives at H and at all of the other vertices the total number of units entering is the same as the total number leaving.

- This diagram showing how the maximum flow can be achieved is not necessarily unique. If we use this method of considering the separate routes but start from the bottom route, ADH, we can arrive at the following arrangement for delivering the maximum flow. Note that the maximum flow is still 680 units but the way it is achieved is different.

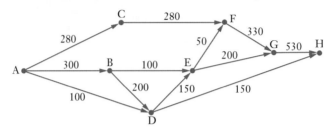

- Considering separate routes, one after another, will not necessarily guarantee that we obtain the maximum flow.

For example, in the network on the right, if we were to consider route ABDC first and direct 40 units along this route, we obtain a total flow of just 40 units, which is clearly not the maximum. To direct 40 units along ABDC in this way is clearly not the wisest thing to do but in some more complicated networks the 'wisest' strategies may not be so obvious. However, the method outlined here, with a little common sense applied, should give the maximum flow for most networks you are likely to meet in this course. Later in this chapter, we will see a way of checking our value to see if it is the maximum flow.

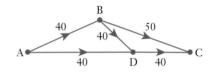

EXAMPLE 1

Determine the maximum flow possible for the network of pipes shown on the right with point A as the source and point F the sink.

Redraw the network showing the flow you would direct along each pipe in order to achieve the maximum flow at F.

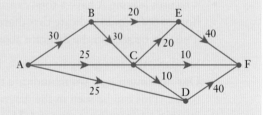

Solution

Using the systematic approach:

ABEF:	20 units	
ABCEF:	10 units	
ACEF:	10 units	
ACF:	10 units	
ACDF:	5 units	
ADF:	25 units	Total: 80 units

The maximum flow from A to F is 80 units and can be achieved as follows:

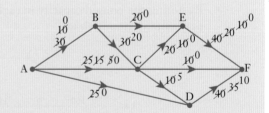

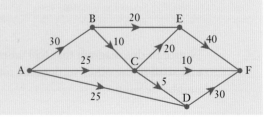

Exercise 6A

For each of the networks shown in questions **1** to **4** write down the maximum flow from the source at A to the sink at X (any working should be done mentally).

1

2

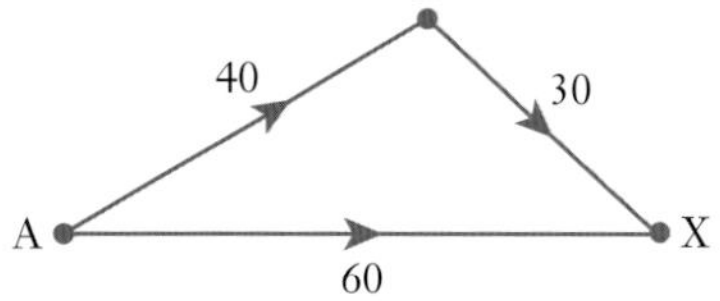

3

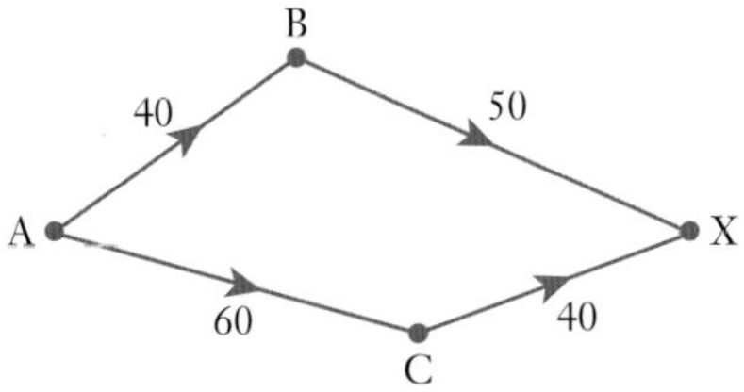

4

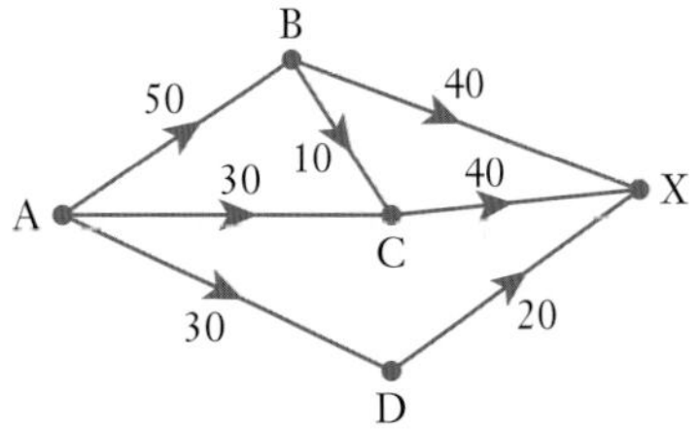

For each of the networks shown in questions **5** to **10** use a systematic approach to determine the maximum flow from the source at A to the sink at Z.

5

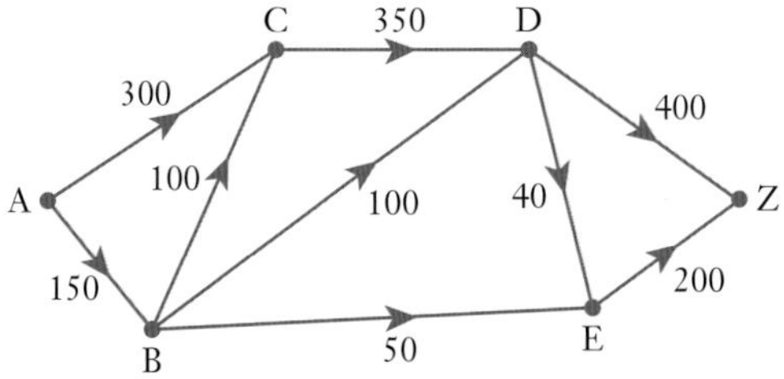

6

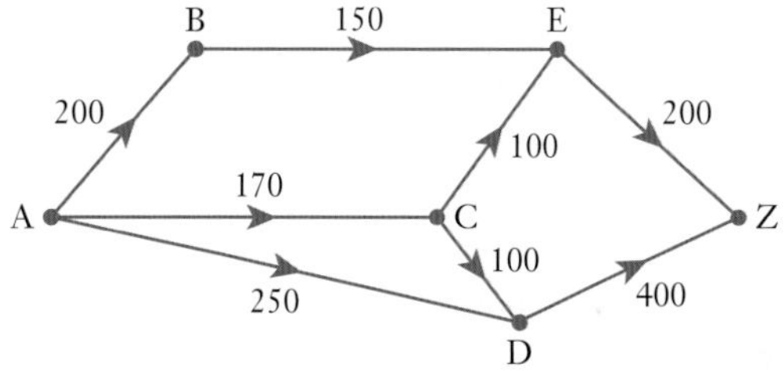

7

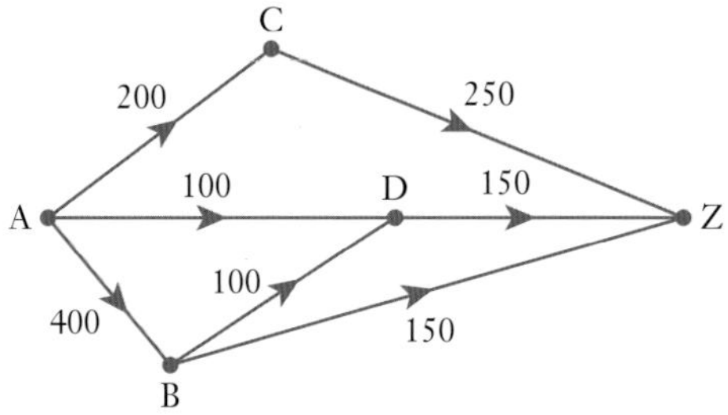

8

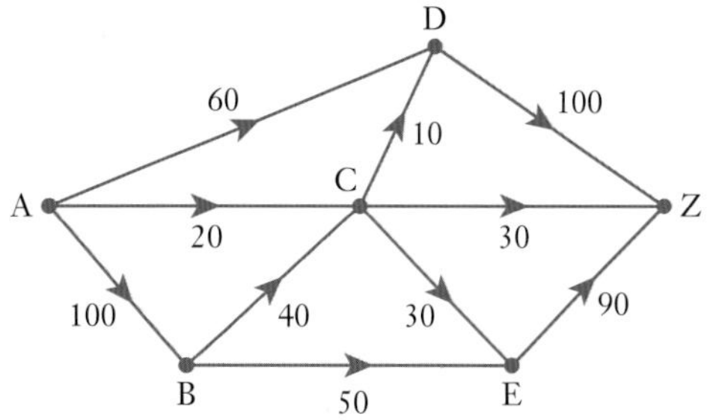

9

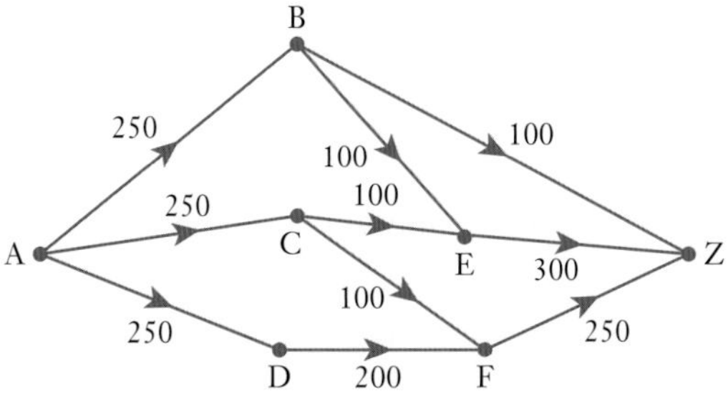

10

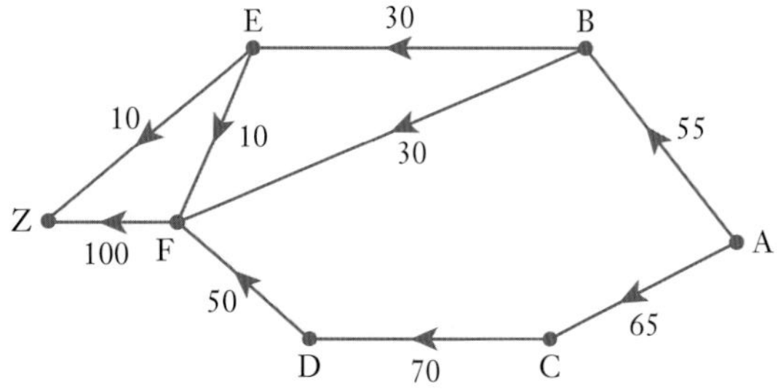

11 For the network shown on the right determine the maximum flow possible from the source at P to the sink at V.

Redraw the network showing the flow you would direct along each arc in order to achieve the maximum flow at V.

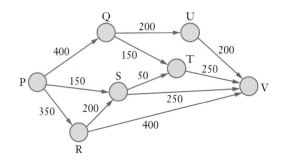

12 Given that the network shown on the right shows the flow along each arc that would achieve the maximum flow from the source at A to the sink at G, determine the values of v, w, x, y and z.

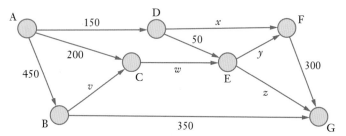

13 The network shown on the right represents a telecommunications system linking five locations, A to E. The number on each edge represents the maximum number of connections that can be active along that edge simultaneously. What is the maximum number of simultaneous connections that can exist from location A to location E?

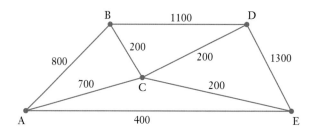

Draw a diagram showing the number of connections you would make along each edge, and show the directions of those connections, in order to produce this maximum situation.

14 The network shown below represents a system of water pipes connected to a main supply at A. The arcs in the network represent the pipes in the system and the numbers on the arcs give the maximum capacity of that section of pipe in units per minute.

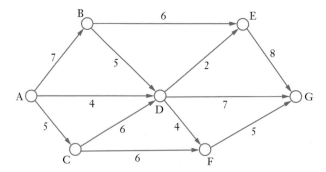

What is the maximum amount of water that could be delivered at point G per minute given that the only restriction is the capacity of the pipes?

One of the three 5-units-per-minute pipes is to be upgraded to an 8-units-per-minute pipe. What would be the effect of each of these three possible upgrades on the maximum flow at G?

Maximum flow = Minimum cut

Earlier it was shown that by choosing an inappropriate route to start with, the total flow we end up with may not be the maximum flow possible. If the network is very complicated, mistakes can easily be made. With careful use of our systematic method it will give us the maximum flow for the networks we are likely to meet in this course but we can check that it is the maximum flow by considering 'cuts'.

Consider the network shown on the right.

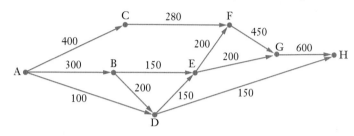

First draw vertical lines through the source on the left and through the sink on the right. A **cut** is then made by starting above the network and drawing a line that cuts only arcs of the network and finishes below the network. (The cut disconnects source from sink.)

The value, or capacity, of the cut is found by summing the cut arcs.

The diagram on the right shows two such cuts, I and II, and their values.

The next diagram shows two more cuts, III and IV.

Note that IV cuts EF from 'below to above' (imagine EF horizontal with its direction of flow being from left to right, i.e. source to sink). This is not included when determining the value of the cut. (The value comes from those arcs that are cut from 'above to below'.)

We can use the fact (not proved here) that **the maximum flow equals the minimum cut** to check our maximum flow. A cut value and a flow value can only be the same when the cut value is a 'min' and the flow value a 'max'. Hence, if we can find a cut equal to our maximum flow then our value must indeed be the maximum.

When we met this network earlier we calculated the maximum flow as 680 units. We can find a cut having this value (see diagram) thus confirming our maximum flow value.

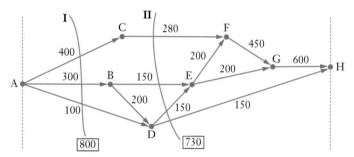

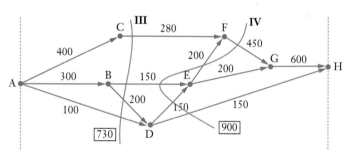

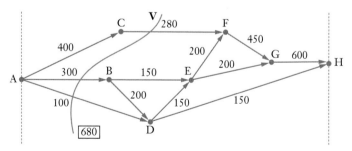

Exercise 6B

1 Determine the value of the cuts I to V shown in the diagram on the right.

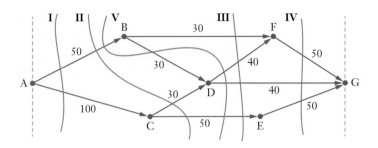

The networks below were all encountered in Exercise 6A. For each network confirm the stated maximum flow by finding a cut of the same value.

2

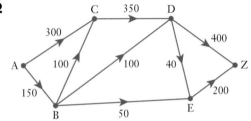

Maximum flow 450 units

3

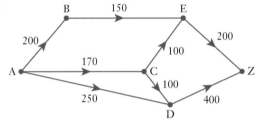

Maximum flow 550 units

4

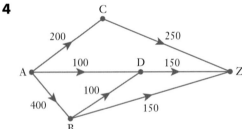

Maximum flow 500 units

5

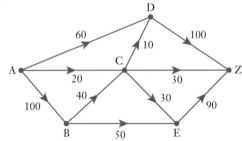

Maximum flow 170 units

6

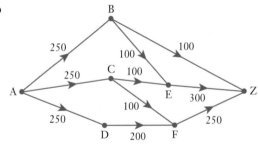

Maximum flow 550 units

7

Maximum flow 100 units

ISBN 9780170395069

Miscellaneous exercise six

This miscellaneous exercise may include questions involving the work of this chapter, the work of any previous chapters, and the ideas mentioned in the Preliminary work section at the beginning of the book.

1 Apply the average (mean) percentage method explained in Chapter 2 (p. 33) to the data given below to determine the seasonal index for each quarter, Q1, Q2, Q3 and Q4, and explain what it is that these indices tell us. Give the indices as percentages rounded to the nearest whole percent.

Year 1					Year 2			
Q1	**Q2**	**Q3**	**Q4**		**Q1**	**Q2**	**Q3**	**Q4**
688	498	248	406		733	632	352	483

2 Find the length of the minimum spanning tree for the network defined by the table on the right.

(A '–' in the table indicates there is no direct route between the locations.)

	A	**B**	**C**	**D**	**E**	**F**	**G**	**H**
A	–	37	70	68	–	80	–	–
B	37	–	40	–	–	–	58	–
C	70	40	–	–	–	–	52	–
D	68	–	–	–	17	–	36	–
E	–	–	–	17	–	20	–	67
F	80	–	–	–	20	–	–	78
G	–	58	52	36	–	–	–	39
H	–	–	–	–	67	78	39	–

3 (Use a calculator with a built-in facility to calculate compound interest calculations or a suitable online finance calculator.)

A loan of $500 000 is taken out to finance the purchase of a house. Interest is charged at 8.5% per annum with interest compounded monthly.

a How much must be repaid per month if the loan is to be paid off in 25 years?

b What is the total amount of interest paid in the 25 years? (Answer to the nearest $10.)

4 A 'one-off' investment to open a savings account, in which interest is calculated annually, is such that if n years later the value of the account is $\$T_n$ then the value one year after that, $\$T_{n+1}$, is such that

$$T_{n+1} = T_n \times 1.072 \quad \text{and} \quad T_0 = 12\,000.$$

a What is the initial investment?

b What is the annual interest rate?

c Using a recursive approach and *not* the financial capability of some calculators and computer programs, determine the value of the account after six years.

5 Under a reducing balance (fixed percentage) depreciation scheme of 12% per annum, how much does an asset initially worth $120 000 depreciate in each of its first three years?

6 The diagram on the right shows the pipe system connecting a number of watering points on a farm to the main source, A. The number on each section indicates the maximum flow that pipe can manage in litres per minute.

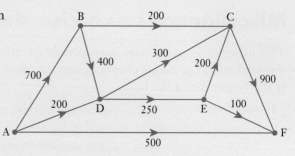

What is the maximum number of litres per minute that could be delivered at outlet F under this system? Make a copy of the diagram and indicate the flow along each pipe (in L/min) to achieve this maximum delivery at F.

7 The following table shows the number of short-term visitor arrivals into Australia for each four-month period from Jan-Feb-Mar-Apr 2008 to Sept-Oct-Nov-Dec 2013.

n	Months	Year	Number of short-term visitors	3-pt MA (M)
1	Jan-Feb-Mar-Apr	2008	1 898 800	–
2	May-Jun-Jul-Aug	2008	1 691 100	
3	Sept-Oct-Nov-Dec	2008	1 922 500	
4	Jan-Feb-Mar-Apr	2009	1 872 200	
5	May-Jun-Jul-Aug	2009	1 587 200	
6	Sept-Oct-Nov-Dec	2009	2 030 800	
7	Jan-Feb-Mar-Apr	2010	1 933 600	
8	May-Jun-Jul-Aug	2010	1 720 400	
9	Sept-Oct-Nov-Dec	2010	2 136 100	
10	Jan-Feb-Mar-Apr	2011	1 959 800	
11	May-Jun-Jul-Aug	2011	1 717 700	
12	Sept-Oct-Nov-Dec	2011	2 093 300	
13	Jan-Feb-Mar-Apr	2012	2 023 100	
14	May-Jun-Jul-Aug	2012	1 777 600	
15	Sept-Oct-Nov-Dec	2012	2 231 500	
16	Jan-Feb-Mar-Apr	2013	2 103 100	
17	May-Jun-Jul-Aug	2013	1 895 700	
18	Sept-Oct-Nov-Dec	2013	2 381 700	–

[Source of data: Australian Bureau of Statistics.]

With the assistance of a computer spreadsheet, or otherwise

a Determine the entries for the final column, the 3-point moving averages, giving values to the nearest hundred.

b Produce a graph with n on the horizontal axis and displaying both the raw data and the moving average data as time series line graphs.

c Use linear regression to obtain the equation of the line of best fit for predicting M, the moving average, given n, i.e. the equation $M = an + b$. (Give a to the nearest 10 and b to the nearest 1000.)

d Interpret the value of 'a' in the equation from part **c**.

ISBN 9780170395069

Project networks

- Project networks
- Constructing a project network
- Miscellaneous exercise seven

Situation

A company is contracted to carry out a particular job. The company splits the job into ten smaller tasks and sub-contracts each of these to other companies and individuals. The main company then monitors the progress of the job by attempting to keep it to the program shown in the network below.

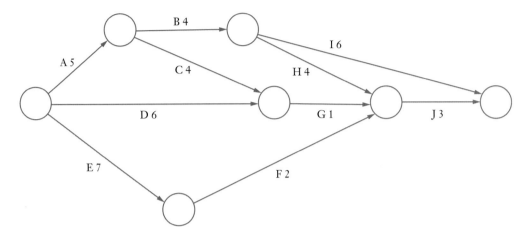

- In this network the ten tasks are labelled A to J and are represented by the directed edges, or arcs. The number on each arc shows the time in days that is allowed for the task.

- The network indicates any task that must be completed before another can commence. Thus task H, for example, has 4 days allowed for it and requires task B to be completed before it can be commenced (and task B required A to be completed before it could be commenced). Provided these 'prior task commitments' are met other tasks can be going on at the same time. For example tasks A, D and E can all be going on at once.

What is the least number of days required for the job to be completed according to the time allocations on the above schedule?

Poor weather may cause task C to take 6 days rather than 4. If this occurs what will be the least number of days for the job now?

If task I were to be completed in 5 days rather than 6, and all others were to be completed using the allotted time, what will be the least number of days for the job now?

If task J were to be completed in 1 day rather than 3, and all others were to be completed using the allotted time, what will be the least number of days for the job now?

WS

Calculating EST and LST

WS

Critical paths, critical times and activity float times

Project networks

The situation on the previous page involved a project network. How did you get on? Did you develop a systematic approach? Did you manage to think it through and answer the questions or did all the things you had to consider leave you confused?

One systematic approach involves two stages:

1. **Forward scan**. I.e. follow the network in the direction of the arrows and at each vertex record the least time needed to reach this stage in the project with all prior tasks completed. In this way we consider the **earliest start time** (EST) of an activity.

2. **Backward scan**. I.e. backtrack through the network recording the latest time each vertex could be reached without delaying the minimum completion time. In this way we consider the **latest start time** (LST) of an activity.

This approach is demonstrated in the following example.

EXAMPLE 1

Find the **minimum completion time** and the **critical path** for the project network shown below. The arcs P, Q, ..., Y represent the tasks involved and the numbers on the arcs indicate the time in hours that the task requires.

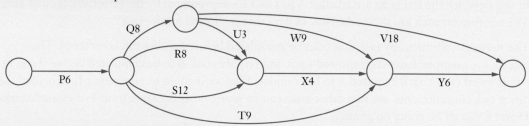

What is the maximum number of extra hours that could be allocated to task S, over and above the 12 in the network, without extending the minimum completion time?

Solution

Work through the network from the start and at each vertex record the least time needed to reach that stage in the project with all prior tasks completed. Check that you understand the placement of the numbers in the circles below.

Earliest this stage can be reached is 6 hours into the project, to allow task P to be finished.

Earliest this stage can be reached is 14 hours into the project, to allow prior tasks (Q and P) to be finished.

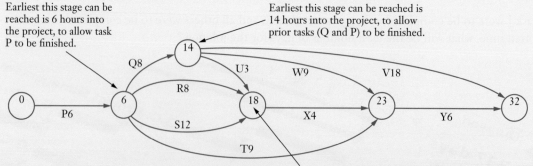

Earliest this stage can be reached is 18 hours into the project, to allow prior tasks U, R and S (and hence Q and P as well) to be finished. The **critical path** to this stage will be P-S because that requires the full 18 hours. Other paths require less and thus have some 'room for delay' (P-R requires 14, P-Q-U requires 17). This 'room for delay' is known as **float** or **slack**.

Thus the **minimum completion time** is 32 hours.

Now, to identify the slackness in the system, work back through the network and at each vertex record in the lower half of the circle the latest time we could reach that stage whilst still being able to complete the project in the minimum completion time. Check that you understand the placement of these numbers in the bottom halves of the circles below.

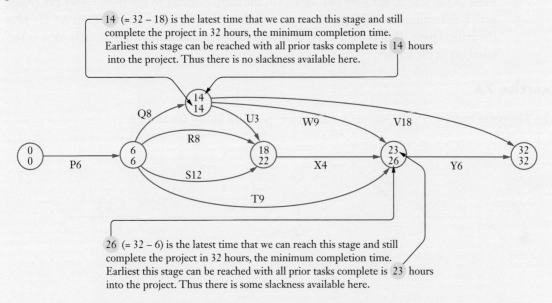

14 (= 32 – 18) is the latest time that we can reach this stage and still complete the project in 32 hours, the minimum completion time. Earliest this stage can be reached with all prior tasks complete is 14 hours into the project. Thus there is no slackness available here.

26 (= 32 – 6) is the latest time that we can reach this stage and still complete the project in 32 hours, the minimum completion time. Earliest this stage can be reached with all prior tasks complete is 23 hours into the project. Thus there is some slackness available here.

Any vertex for which the earliest time the vertex can be reached is the same as the latest time the vertex can be reached lies on the **critical path**. If we journey from start to finish, through such vertices we can obtain the critical path, P-Q-V. If any of these tasks are delayed the minimum completion time will be extended.

Consider task S and notice that this task will be ready to start 6 hours into the project and does not have to finish until 22 hours into the project. Thus we could allow 16 hours for this 12-hour task. Thus we could allocate an extra 4 hours for task S, over and above the 12 in the network. Task S is said to have a **float**, or **slack** time of 4 hours.

Consider task X. We say that X has an **earliest start time** (EST) of 18 hours and a **latest start time** (LST) of 22 hours (26 – 4). X has 4 hours float.

Consider task W. It has an **earliest start time** (EST) of 14 hours and a **latest start time** (LST) of 17 hours (26 – 9). W has 3 hours float.

All activities on the critical path have zero float time.

Note: • In the above example the diagram is drawn twice and comments are included in boxes. The reader would not have to include such comments and would only need one diagram.

• The float, or slack, considered here is the time by which an activity can be delayed, from its early start time, without altering the project's minimum completion time. It is sometimes referred to as the **total float** of an activity.

• Consider the diagram:

The numbers in the circles may at first glance give the impression that there is no float, or slack, in the system but with two arcs from the second vertex to the third vertex the critical path is determined by taking task C, the task that takes the greater time. Thus A - C - D is the critical path. Task B has 2 units of float, a fact that can be missed if we only look at the numbers in the circles.

Exercise 7A

1 The times for tasks P to Z shown in the project network below are in days.

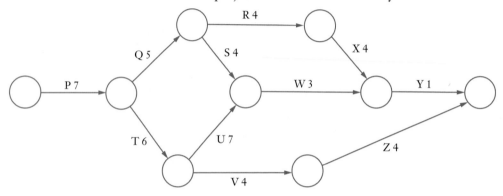

a Determine the minimum completion time.

b Determine the critical path.

c Task V is a task that is expected to take 4 days. Assuming all the other tasks can be completed in exactly the stated times, how many extra days could be allocated to task V without altering the minimum completion time?

2 The times for tasks A to H shown in the project network below are in hours.

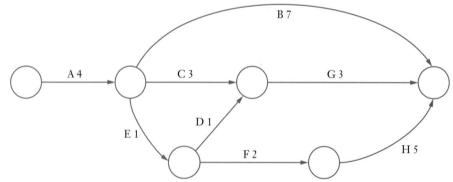

a Determine the minimum completion time.

b Determine the critical path.

c Some problems are experienced with task D and it requires an extra 3 hours. What is the minimum completion time for the project now?

d How many extra hours could be allowed for task B, over and above the 7 hours allocated on the above network, without extending the minimum completion time?

ISBN 9780170395069

3 The times for the tasks shown in the project network below are in days.

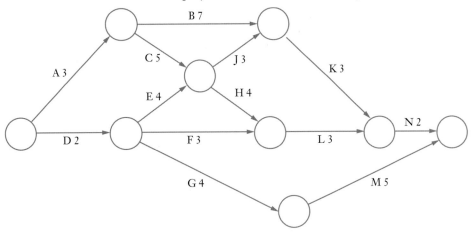

a Determine the minimum completion time.

b Determine the critical path.

c Due to poor weather, tasks D and B each suffer a two day delay in commencement. What is the minimum completion time now?

4 The times for the tasks P to Z shown in the project network below are in hours.

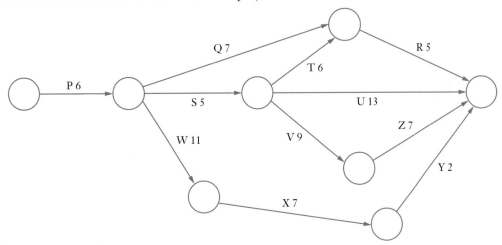

a Determine the minimum completion time.

b Determine the critical path.

c The project managers can employ extra staff and reduce the time required for task V by 3 hours. If they take this option will the minimum completion time be cut by 3 hours?

d Instead of the option outlined in part **c** the managers decide to allocate extra staff to *one* particular task and this reduces the time required for that task by 2 hours and it cuts the minimum completion time by 2 hours. Which task did they allocate the extra staff to?

5 The times shown in the project network below are in days.

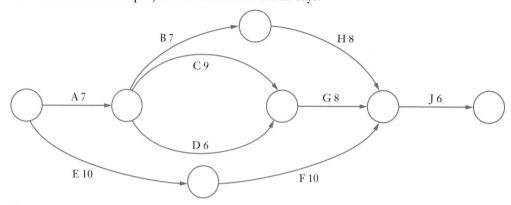

a Determine the minimum completion time.

b Determine the critical path.

c What is the earliest start time (EST) for task H?

d What is the latest start time (LST) for task H if the minimum completion time is to be unchanged from that given in the answer to part **a**?

e What is the float time for task H?

f What is the float time for task F?

g What is the float time for task D?

h Due to staff sickness it is necessary to reallocate the staff involved on tasks D and F. This will cause one of these jobs to require 5 extra days to complete. Which of the two tasks should be the one chosen to require this extra time if the completion time is to be kept as low as possible?

6 The times for tasks P to Z shown in the project network below are in hours.

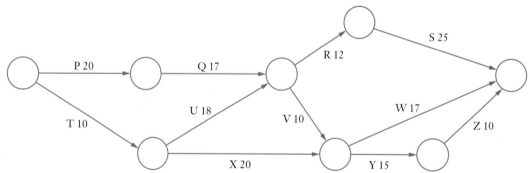

a Determine the minimum completion time.

b Determine the critical path.

c What is the earliest start time after the commencement of the project that task X can commence?

d What is the latest start time after the commencement of the project that task X can commence without altering the minimum completion from **a**?

e What is the slack time for task X?

f What is the latest time after the commencement of the project that task W can commence without altering the minimum completion from **a**?

g What is the slack time for task W?

7 The times for tasks P to Z shown in the network below are in minutes.

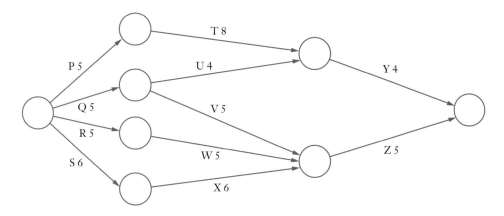

a Determine the minimum completion time and critical path(s) for the above project network.

b How much slack time do activities Q and R have?

8 The times for the tasks shown in the project network below are in days.

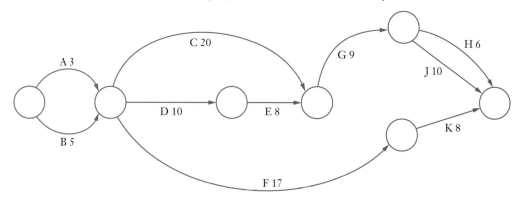

a Determine the minimum completion time.

b Determine the critical path.

c What is the maximum number of days the commencement of task F could be delayed without altering the minimum completion time?

d What float time does activity E have?

e What float time does activity H have?

9 The project network below is for the repair of a car at a garage.

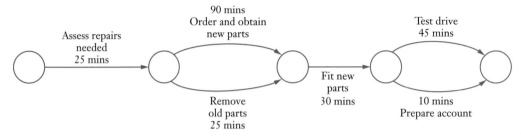

a Determine the minimum completion time.

b The owner of the car delivers it to the garage at 9 o'clock one morning. Determine the earliest time it will be ready, assuming the time for each task is as in the network.

c The new parts are in fact all in stock in the garage and do not need to be ordered from elsewhere. This task then takes just 5 minutes rather than the 90 allowed in the network. What is the earliest completion time now, given that work commences at 9 a.m.?

10 A cook prepares a meal consisting of casserole, dumplings, fruit salad and chilled wine according to the project network shown below.

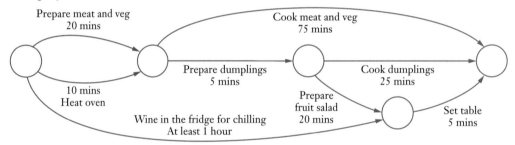

a If the meal has to be ready for 6 p.m. what is the latest time that the first activity can commence?

b What is the latest time that the warm wine can be placed in the fridge for chilling?

c With the wine in the fridge and just as she puts the meat and veg in the oven to cook and is about to start preparing the dumplings, the cook's phone rings. What is the maximum time she can spend on the phone without delaying the completion time? (Assume that she cannot prepare the dumplings and be on the phone at the same time.)

ISBN 9780170395069

Constructing a project network

The previous exercise involved questions being asked about given project networks. In some cases we may not be given the project network but instead be given sufficient information to allow the network to be constructed. Examples 2 and 3 are of this type.

Construct a network for a project consisting of activities A to H listed below.

Activity	Time	Order
A	7 hours	• Activities A and B start together.
B	9 hours	
C	8 hours	• Activities C and D can both commence when A is complete.
D	3 hours	
E	7 hours	• Activities F and E can both commence provided both B and D have finished.
F	4 hours	
G	5 hours	• Activity G can commence provided C and F are both complete.
H	3 hours	• Activity H can commence provided E and G are both complete.

Solution

Activities A and B start together:

Activities C and D can both commence when A is complete:

Activities F and E can both commence provided both B and D have finished:

Activity G can commence provided C and F are both complete:

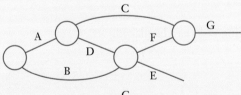

Activity H can commence provided E and G are both complete:

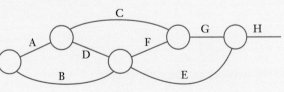

Complete the project network by adding the final circle, the arrows and the activity times:

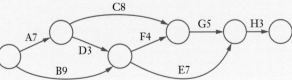

EXAMPLE 3

Construct a project network for a project that consists of the tasks shown on the right.

Task	Time	Immediate predecessors
A	7 days	–
B	4 days	A
C	10 days	B
D	3 days	B
E	4 days	B
F	6 days	A
G	3 days	F
H	7 days	F
I	5 days	E, G
J	4 days	D, H, I

Solution

Start the network with those tasks that have no predecessors, in this case task A:

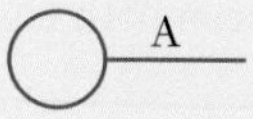

Tasks B and F have A as immediate predecessor:

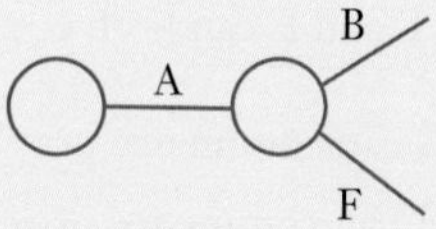

With B in place tasks C, D and E follow:

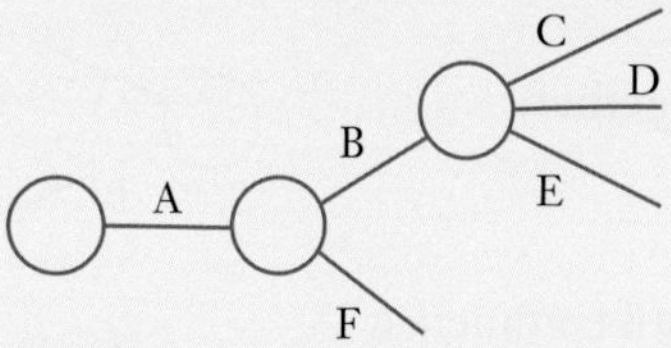

Tasks G and H can be placed:

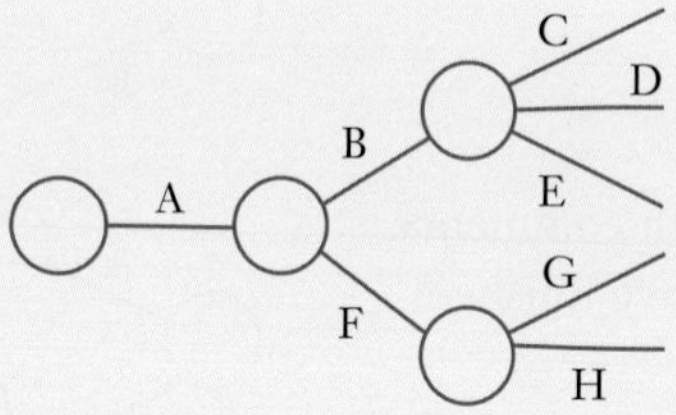

The remaining tasks can be placed:

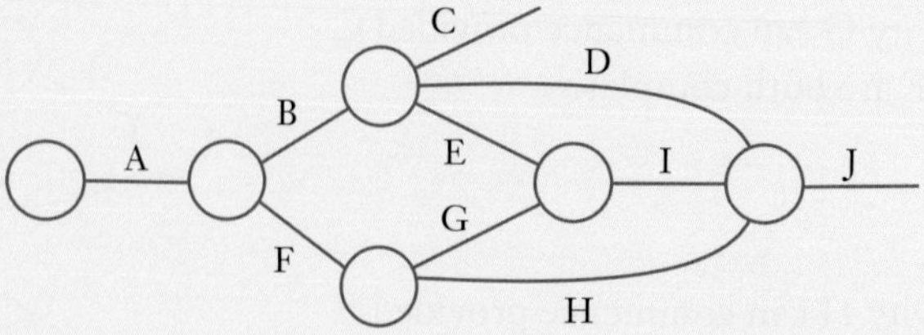

And the network can be completed:

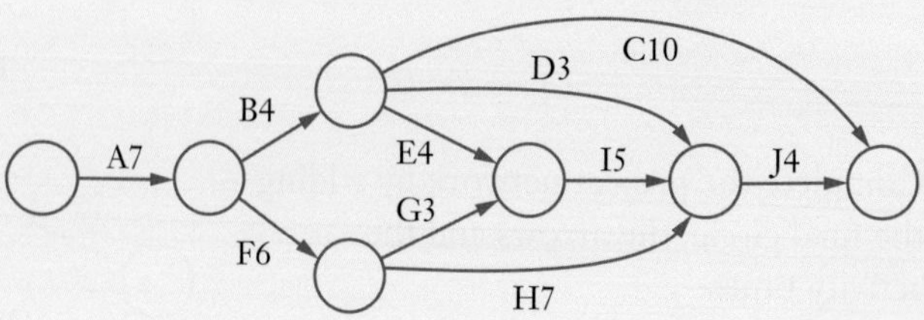

Exercise 7B

1 a Construct a project network for a project consisting of tasks P to X.

b Determine the minimum completion time and the critical path.

c How many weeks over the scheduled time can task U take without altering the minimum completion time? (Assume all other tasks are completed as planned.)

Task	Scheduled time	Immediate predecessors
P	7 weeks	–
Q	5 weeks	–
R	3 weeks	P
S	4 weeks	P
T	6 weeks	–
U	9 weeks	T
V	5 weeks	S, Q
W	7 weeks	R
X	2 weeks	V, W

2 a Construct a network for a project consisting of the tasks shown in the table.

b Determine the minimum completion time and the critical path.

c Task V is subject to delay and in fact takes 12 days to complete, not 7. What is the minimum completion time now?

Task	Time	Immediate predecessors
Q	6 days	–
R	8 days	–
S	5 days	Q
T	4 days	Q
U	3 days	R
V	7 days	R
W	2 days	T, U
X	4 days	S, W
Y	3 days	V, Z
Z	5 days	S, W

3 a Construct a project network for a project that consists of the tasks shown in the table.

b Determine the minimum completion time and the critical path.

c By how many minutes can activity E be delayed without altering the minimum completion time?

Task	Time	Immediate predecessors
A	16 minutes	–
B	21 minutes	A
C	28 minutes	–
D	25 minutes	B, C
E	5 minutes	B, C
F	25 minutes	E
G	16 minutes	D, H
H	17 minutes	E

4 a Construct a project network for a project consisting of activities A to G.

b Determine the minimum completion time and the critical path.

c Lack of harmony on the industrial front threatens to delay task F. If all the other tasks can be completed according to the times given, how many extra days can be allowed for activity F, over and above the scheduled 11, without increasing the minimum completion time?

Activity	Time	Order
A	10 days	• Activities C and D start together.
B	21 days	• Activities E and F must be after activity C.
C	16 days	
D	35 days	• Activities A and B need activity E to be finished before they can start.
E	15 days	
F	11 days	• Activity G can commence only after F is complete.
G	17 days	

5 a Construct a project network for a project consisting of activities A to I.

b Determine the minimum completion time and the critical path.

c Which of the three tasks that start together, A, D and H, could experience delays causing them to be completed in a time greater than that stated yet still not alter the minimum completion time for the project?

Activity	Time	Order
A	6 hours	• Activities A, D and H start together.
B	7 hours	
C	4 hours	• Activity E must be after activity A.
D	7 hours	• Activities B, F and I must be after D.
E	10 hours	
F	3 hours	• Activity C can commence only after both I and H are complete.
G	5 hours	
H	9 hours	• Activity G can commence only after both E and F are complete.
I	4 hours	

6 Construct a project network for a project consisting of tasks P to Z as listed in the table.

Determine the minimum completion time and the critical path.

Activity	Duration	Requirements
P	5 minutes	• Q, T and W start when P is completed.
Q	6 minutes	
R	10 minutes	• S can start when T and Y are completed.
S	2 minutes	
T	4 minutes	• X can start when W is completed.
U	9 minutes	• R and Y can start when Q is completed.
V	4 minutes	
W	3 minutes	• Z can start when both S and X are completed.
X	10 minutes	
Y	4 minutes	• U can start when both S and X are completed.
Z	4 minutes	• V can start when R and Z are completed.

ISBN 9780170395069

7 a Construct a project network for a project that consists of tasks R to Z as listed in the table.

b Determine the minimum completion time and the critical path.

c It is thought that activities U, V and Y may *each* require 2 extra days for completion than is given in the table. What would be the minimum completion time if all three of these tasks do require this extra time (and all other tasks have the duration shown in the table).

Activity	Duration	Order
R	16 days	• R and T start together.
S	22 days	• S, U, V start once both R and T are finished.
T	25 days	
U	20 days	• W starts when S is finished.
V	28 days	• X starts when W is finished.
W	17 days	• Y starts when U is finished.
X	15 days	• Z starts when both V and Y are finished.
Y	13 days	
Z	18 days	

8 A company decides to interview three of the many applicants for a job. The three people are Mr James, Ms Kerry and Ms Maine.

The interview process will involve the following tasks:

Activity	Description	Time (mins)	Order
A	Three applicants to meet informally with interview panel for coffee	30	The order in which these activities should be completed is given below.
B	Explain interview procedure to the applicants as a group	5	• Activity A is the first activity and is followed by B.
C	Interview Mr James	45	• Activities C, D and E all follow B.
D	Ms Kerry does written test	30	• Activities F, G and H occur once all of C, D and E have been completed.
E	Ms Maine tours work site	30	
F	Interview Ms Maine	45	• Activities I, J and K occur once all of F, G and H have been completed.
G	Mr James does written test	30	
H	Ms Kerry tours work site	30	• Activity L follows I, J and K.
I	Interview Ms Kerry	45	• Activity M follows L.
J	Ms Maine does written test	30	
K	Mr James tours work site	30	
L	Interview panel meet to make decision	45	
M	Panel chairman informs applicants of panel's decision	2	

Display the process as a project network.

If activity A, the informal meeting over coffee, finishes at 9.30 a.m. when does Ms Kerry finish her interview?

9 a Construct a project network for a project consisting of activities Q to Z.

Activity	Time	Order
Q	4 hours	• Q, S and T can all start without needing other activities beforehand.
R	4 hours	
S	7 hours	
T	9 hours	• R can commence provided Q has finished.
U	3 hours	• U and V can commence provided R, S and T have all finished.
V	11 hours	
W	14 hours	
X	7 hours	• W, X and Y can commence provided U has finished.
Y	9 hours	
Z	3 hours	• Z can commence provided V, X and Y have all finished.

b Determine the minimum completion time and the critical path.

c Determine how many hours activity Z could run over its allocated time and still not alter the minimum completion time if

 i all the other tasks take exactly their allocated times,

 ii task Y is completed in 7 hours rather than the 9 allowed for.

10 A printing company has the job of printing a book for a client. Once they receive the original pages from the client they carry out the following activities:

Activity	Description	Time (days)
A	Obtain necessary materials (paper, card etc.)	3
B	Paste up originals to form masters	3
C	Make printing plates from masters	1
D	Print pages	4
E	Print cover	1
F	Fold and collate pages	2
G	Bind pages	3
H	Attach cover	2
I	Deliver to client	1

The order in which these activities should be completed is given below.

• Obtaining necessary materials and pasting up can start together.

• Making the plates can only be done once the paste up is complete.

• The printing of the pages and the cover can both commence provided both the materials are obtained and the plates are made.

• The pages can be folded and collated as soon as they are all printed.

• Page binding can occur following folding and collation.

• The printed covers are attached once page binding is complete.

• When all complete the books are delivered to the client.

ISBN 9780170395069

a Construct a project network for this job.

b Determine the minimum completion time and list the tasks that form the critical path.

c The covers are about to be printed when the machine that prints them breaks down and a new part has to be sent from Germany. What is the longest time this machine can be out of use without altering the minimum completion time?

d The company is considering purchasing a new machine that will fold, collate and bind the pages in a total of 4 days. By how much would this new machine reduce the minimum completion time?

11 An airline company is keen to reduce the time its planes are on the ground at a particular airport because of the high fees the airport authority charges. The company is charged for each minute that it has an aircraft at an airport terminal.

Construct a project network for the 'on ground' tasks listed below and determine the minimum completion time.

iStock.com/mbbirdy

Activity	Description	Time (mins)	Preceded by
A	Support service vehicles in position	5	–
B	Passengers disembark	15	A
C	Luggage off	8	A
D	Old food trolleys off	5	A
E	Refuel	10	A
F	Security check of empty aircraft	15	B, C
G	New passengers on	25	F
H	Luggage on	12	F
I	Food trolleys on	5	D
J	Remove old waste disposal units	8	A
K	Fit new waste disposal units	5	J

Which tasks should they attempt to speed up if they are to reduce the time at the terminal?

12 Though generally avoided in this text, the following situation is one requiring the introduction of a 'dummy activity' of zero weight for a network diagram to be formed. Use this idea to represent the following as a project network with dummy activity Q, and determine the critical path and shortest completion time.

• Tasks A (5 hours) and B (4 hours) start together.

• Task C (7 hours) can commence when both A and B are completed.

• Task D (6 hours) can commence when task A is completed.

13 In order to keep the production of a school yearbook on time the organising committee list the tasks involved and the order in which they must occur.

Task	Activity	Time (weeks)	Order
A	Advertise fact that articles are required and receive articles	10	• Tasks A, B and D start together.
B	Take individual photographs of all year twelve students	3	• Task C can commence when task B finishes.
C	Check all individual photographs and retake as necessary	1	• Task E can commence when task D finishes.
D	Get personal details such as nickname etc. from each year 12	3	• Task F can commence when task A finishes.
E	Have personal details typed to same style and format	1	• Task G can commence when task F finishes.
F	Select articles to be included from those submitted	1	• Task H can commence when task G finishes.
G	Edit selected articles for grammar and acceptability	1	• Task I can commence when tasks C, E and H are finished.
H	Have all selected articles typed to same style and format	1	• Task J can commence when all other tasks have been completed.
I	Paste up articles, personal details and photos as ordered pages	3	
J	Print yearbook	2	

Represent this information as a project network and determine the minimum completion time.

Miscellaneous exercise seven

This miscellaneous exercise may include questions involving the work of this chapter, the work of any previous chapters, and the ideas mentioned in the Preliminary work section at the beginning of the book.

1 Explain what it means to say that the seasonal index for January is 1.12.

2 A loan for $280 000 is taken out with compound interest charged at a rate of 10% per annum compounded monthly. If repayments of $2800 are made at intervals of one month after the start of the loan, how much is owed five years after its commencement?

3 An analysis of the number of cars a particular manufacturer sells in each month, over a number of years, gives the seasonal index for January as 92% and for February as 88%.

The following year the manufacturer sells 2031 cars in January and 1997 cars in February. Express the number of cars sold in these two months as seasonally adjusted numbers.

ISBN 9780170395069

4 Which of the following graphs show
   a an increasing trend?
   b a decreasing trend?
   c data suitable for linear modelling?
   d data with irregular fluctuations?
   e data with seasonal fluctuations?

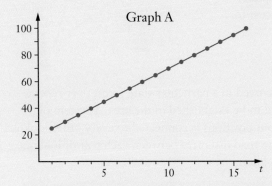

Graph A

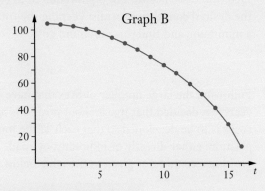

Graph B

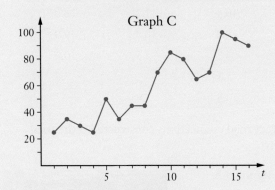

Graph C

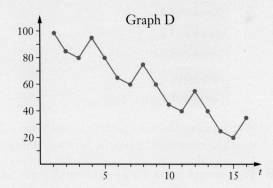

Graph D

5 Determine the maximum number of units that can be delivered at the sink, G, from the source, A, for the network shown below.

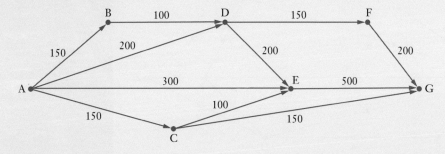

If the capacity of arc AC could be increased from the 150 units shown in the diagram to 250 units how would this alter the maximum number of units that can be delivered from A to G?

6 The network on the right shows four proposed off-shore drilling wells and the land-based processing base. Pipelines are to be laid connecting each well to the base, either directly or via one or more of the other wells. The cost of each line is shown in $1000s.

Determine the network of pipes required to complete the desired connections whilst keeping the total cost to a minimum, and find this minimum cost.

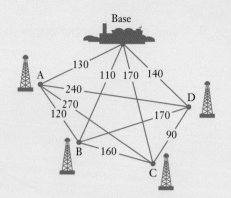

7 Following the large number of fires that have occurred in a particular state forest over recent years it is decided that five lookout positions need to be established in the area. A system of roads is to be developed so that each of the lookout positions is connected to every other lookout position, either directly or indirectly, by road. The road distances between each pair of lookout positions would be as given in the table below, in kilometres.

	Lookout 1	Lookout 2	Lookout 3	Lookout 4	Lookout 5
Lookout 1	–	5.0	10.5	11.2	6.0
Lookout 2	5.0	–	7.0	10.0	9.0
Lookout 3	10.5	7.0	–	5.5	10.8
Lookout 4	11.2	10.0	5.5	–	7.8
Lookout 5	6.0	9.0	10.8	7.8	–

Which pairs of lookout positions should have direct road connections constructed if the total length of construction is to be kept to a minimum?

8 The network below shows a system of roads linking town A to town Z.

Points B, C, D, E and F are road junctions.

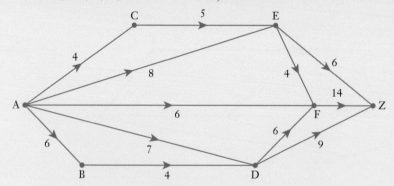

The number on each road gives the maximum number of vehicles (in hundreds) that can travel along that section of road, in the direction indicated, per hour.

Determine the greatest number of vehicles that could arrive at Z per hour, having come from A.

ISBN 9780170395069

Assignment
problems

Situation

In swimming competitions, a *4 × 100 metre medley relay* involves four different swimmers, each swimming 100 metres and with each doing a different stroke.

The first swimmer swims 100 metres backstroke,
the second swimmer swims 100 metres breaststroke,
the third swimmer swims 100 metres butterfly,
and the fourth swimmer swims 100 metres freestyle.

A swimming club has four swimmers who will form the team for the men's 4 × 100 metres medley relay team. They are Alex, Ben, Chris and Devon. The season personal best times for each of these swimmers, in each of the strokes, over 100 metres are as given in the table below.

(Times are given as minutes.seconds.hundredths-of-a-second. Thus 1.08.72 means 1 minute, 8 seconds and 72 hundredths of a second, i.e. 68.72 seconds all together.)

	Backstroke	Breaststroke	Butterfly	Freestyle
Alex	1.08.72	1.20.17	56.28	58.17
Ben	1.01.24	1.09.21	1.10.21	53.05
Chris	1.03.21	1.17.16	1.03.27	55.21
Devon	1.04.17	1.08.71	1.17.24	54.37

Which swimmer would you have swimming each leg of the relay to give the fastest possible time for the team, based on the season personal best times?

Suppose instead that Alex's season personal best time for the butterfly is 1.08.17. I.e. the table of season personal best times would now be as follows:

	Backstroke	Breaststroke	Butterfly	Freestyle
Alex	1.08.72	1.20.17	1.08.17	58.17
Ben	1.01.24	1.09.21	1.10.21	53.05
Chris	1.03.21	1.17.16	1.03.27	55.21
Devon	1.04.17	1.08.71	1.17.24	54.37

What would be the preferred order now?

iStock.com/cmcderm1

The situation on the previous page required us to assign each swimmer to a stroke in order to give the fastest medley relay team, based on season personal best times. This chapter considers **assignment problems** of this type, also called **allocation problems**. The swimming situation required us to allocate the swimmers in order to give the best total *time*. Not all problems of this type involve optimising time. Some may be concerned with optimising cost or distance travelled etc.

Assignment problems

Suppose a company wishes to use two transport companies, A and B, for two deliveries, 1 and 2, giving one delivery to each transport company. The locations of each pick-up point and delivery point in relation to each transport company, and the transport company costs, means that for each job, each of the transport companies quotes a different price. The quoted prices are as follows:

	Delivery 1	Delivery 2
Company A	$145	$180
Company B	$164	$191

Alternatively, this information could be given as a bipartite graph:

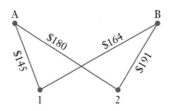

Can you see which allocation of the two jobs will minimise the total cost (remembering that we must give one delivery to each company)?

If we write AB to indicate company A doing delivery 1 and company B doing delivery 2, there are only two allocations that are possible:

$$AB \quad \text{and} \quad BA$$

AB: Company A doing delivery 1 and company B doing delivery 2.

$$\begin{aligned} \text{Total cost} &= \$145 + \$191 \\ &= \$336 \end{aligned}$$

BA: Company B doing delivery 1 and company A doing delivery 2.

$$\begin{aligned} \text{Total cost} &= \$164 + \$180 \\ &= \$344 \end{aligned}$$

Cheapest solution: Company A doing delivery 1 and company B doing delivery 2 for a total cost of $336.

Is this the solution you chose just by considering the table of costs?

Now let us consider a situation involving three companies, A, B and C, and three deliveries, 1, 2 and 3, again with one delivery to be assigned to each company.

Suppose the costs this time are as follows:

	1	2	3
A	$170	$150	$260
B	$160	$160	$210
C	$190	$170	$270

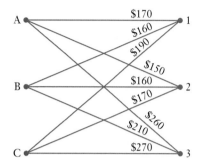

Now, writing ABC to mean company A does delivery 1, B does delivery 2 and C does delivery 3, there are six possible ways the assignments can be organised:

ABC: $170   + $160   + $270   = $600
ACB: $170   + $170   + $210   = $550   ←
BAC: $160   + $150   + $270   = $580
BCA: $160   + $170   + $260   = $590
CAB: $190   + $150   + $210   = $550   ←
CBA: $190   + $160   + $260   = $610

Thus there are two allocations that equally give the minimum cost of $550:

A do delivery 1, B do delivery 3 and C do delivery 2,

or

A do delivery 2, B do delivery 3 and C do delivery 1.

Allocating for a maximum

Some allocation problems may involve situations in which the most desirable, or optimum, situation involves a maximum rather than a minimum. Consider for example allocating three workers to three machines. Each machine makes different components and each worker has different competencies on each machine. However the company wants to allocate each worker to a machine such that the total number of components produced is maximised. Suppose that the number of components produced in a day by each worker, A, B and C, on each machine, 1, 2 and 3, is as follows:

	Machine 1	Machine 2	Machine 3
Worker A	1020	950	1430
Worker B	760	920	1200
Worker C	950	1000	1450

Which worker should be assigned to each machine to maximise the total number of components produced by the three workers?

Try to find a solution before turning the page where a solution is presented. Perhaps set up a spreadsheet showing the various arrangements and their totals.

Writing ABC to mean worker A is on machine 1, B is on machine 2 and C is on machine 3, there are six possible ways the workers can be assigned to machines:

ABC:	1020	+	920	+	1450	=	(3390)	← maximum total.
ACB:	1020	+	1000	+	1200	=	3220	
BAC:	760	+	950	+	1450	=	3160	
BCA:	760	+	1000	+	1430	=	3190	
CAB:	950	+	950	+	1200	=	3100	
CBA:	950	+	920	+	1430	=	3300	

Hence we should assign worker A to machine 1, worker B to machine 2 and worker C to machine 3 for a total number of components produced of 3390.

Four machines, four tasks

The next example returns to an optimum situation involving a minimum but now a 4×4 table is involved.

EXAMPLE 1

Four machines are to be assigned to carry out four tasks, with one machine for each task, and the aim of minimising the total machine time taken to complete the tasks.

The time each machine is estimated to complete each task is given on the right, in hours. Find the allocation of machines to tasks that will minimise total machine time.

	Task 1	Task 2	Task 3	Task 4
Machine A	15	18	9	12
Machine B	13	21	11	15
Machine C	12	19	8	13
Machine D	14	24	10	11

Solution

Writing ABCD to mean machine A is on task 1, machine B is on task 2, machine C is on task 3 and machine D is on task 4, there are 24 possible ways the machines can be assigned to the tasks:

ABCD:	15	+	21	+	8	+	11	=	55	CABD:	12	+	18	+	11	+	11	=	52
ABDC:	15	+	21	+	10	+	13	=	59	CADB:	12	+	18	+	10	+	15	=	55
ACBD:	15	+	19	+	11	+	11	=	56	CBAD:	12	+	21	+	9	+	11	=	53
ACDB:	15	+	19	+	10	+	15	=	59	CBDA:	12	+	21	+	10	+	12	=	55
ADBC:	15	+	24	+	11	+	13	=	63	CDAB:	12	+	24	+	9	+	15	=	60
ADCB:	15	+	24	+	8	+	15	=	62	CDBA:	12	+	24	+	11	+	12	=	59
BACD:	13	+	18	+	8	+	11	=	(50) ←	DABC:	14	+	18	+	11	+	13	=	56
BADC:	13	+	18	+	10	+	13	=	54	DACB:	14	+	18	+	8	+	15	=	55
BCAD:	13	+	19	+	9	+	11	=	52	DBAC:	14	+	21	+	9	+	13	=	57
BCDA:	13	+	19	+	10	+	12	=	54	DBCA:	14	+	21	+	8	+	12	=	55
BDAC:	13	+	24	+	9	+	13	=	59	DCAB:	14	+	19	+	9	+	15	=	57
BDCA:	13	+	24	+	8	+	12	=	57	DCBA:	14	+	19	+	11	+	12	=	56

Assign machine A to task 2, B to task 1, C to task 3 and D to task 4. Total time 50 hours.

Are you able to set up a spreadsheet for the above listing that would be adaptable for any other 'four machine, four task' situation?

ISBN 9780170395069

Exercise 8A

The following tables show the cost that trucking firms A, B, C, … charge to deliver loads 1, 2, 3, … .

Assign trucking firms to loads such that in each case each firm is assigned a different load and the total cost is **minimised**.

1

	Load 1	Load 2
Firm A	$150	$120
Firm B	$120	$110

2

	Load 1	Load 2	Load 3
Firm A	$100	$140	$150
Firm B	$90	$170	$100
Firm C	$110	$140	$130

3

	Load 1	Load 2	Load 3	Load 4
Firm A	$700	$650	$850	$1200
Firm B	$600	$700	$950	$1050
Firm C	$750	$600	$750	$1250
Firm D	$650	$750	$850	$1150

The following tables show the number of components each operator, A, B, C, … can produce in a day when operating machines 1, 2, 3, … .

Assign operators to machines such that in each case each operator is assigned to a different machine and the total number of components produced in a day is **maximised**.

4

	Machine 1	Machine 2
Operator A	80	90
Operator B	85	100

5

	Machine 1	Machine 2	Machine 3
Operator A	210	200	170
Operator B	230	210	190
Operator C	200	180	180

6

	Machine 1	Machine 2	Machine 3	Machine 4
Operator A	60	90	105	80
Operator B	80	90	110	85
Operator C	65	95	110	90
Operator D	75	85	115	75

An algorithm for solving allocation problems

Whilst the process of listing all possibilities to determine the optimum arrangement is a valid process, there is an algorithm we can use to determine the optimum allocation, without having to list all possibilities. The steps of this algorithm are shown below.

Consider the situation of assigning four deliveries to four courier firms with each firm taking one delivery and with the allocation aiming to minimise the total cost.

Suppose the costs charged by each firm are as follows

	Delivery 1	Delivery 2	Delivery 3	Delivery 4
Courier firm A	$250	$100	$140	$120
Courier firm B	$220	$110	$150	$110
Courier firm C	$220	$90	$160	$130
Courier firm D	$260	$140	$170	$120

Step 1:   Identify the **cost matrix**:
$$\begin{bmatrix} 250 & 100 & 140 & 120 \\ 220 & 110 & 150 & 110 \\ 220 & 90 & 160 & 130 \\ 260 & 140 & 170 & 120 \end{bmatrix}$$

Step 2:   Subtract the smallest number in each row from every number in that row.

Hence we subtract   100   from each element of row 1,
　　　　　　　　　110   from each element of row 2,
　　　　　　　　　　90   from each element of row 3,
and　　　　　　　　120   from each element of row 4.

This gives the matrix:
$$\begin{bmatrix} 150 & 0 & 40 & 20 \\ 110 & 0 & 40 & 0 \\ 130 & 0 & 70 & 40 \\ 140 & 20 & 50 & 0 \end{bmatrix}$$

Step 3:   Subtract the smallest number in each column from every entry in that column.

Hence we subtract   110   from each element of column 1,
　　　　　　　　　　0   from each element of column 2,
　　　　　　　　　40   from each element of column 3,
and　　　　　　　　0   from each element of column 4.

This gives the matrix:
$$\begin{bmatrix} 40 & 0 & 0 & 20 \\ 0 & 0 & 0 & 0 \\ 20 & 0 & 30 & 40 \\ 30 & 20 & 10 & 0 \end{bmatrix}$$

Step 4:   Now draw as few **horizontal** and/or **vertical** straight lines as possible through all of the zeros. This question involves a 4 × 4 matrix so if the number of such lines equals 4, a solution can be found from this matrix, as is the case here.
$$\begin{bmatrix} 40 & 0 & 0 & 20 \\ 0 & 0 & 0 & 0 \\ 20 & 0 & 30 & 40 \\ 30 & 20 & 10 & 0 \end{bmatrix}$$

ISBN 9780170395069

Step 5: We now need to choose **one** zero in each column and **one** zero in each row.

(Hint: First consider any rows and columns that only have one zero entry in them as that entry must be chosen.)

$$\begin{bmatrix} 40 & 0 & \boxed{0} & 20 \\ \boxed{0} & 0 & 0 & 0 \\ 20 & \boxed{0} & 30 & 40 \\ 30 & 20 & 10 & \boxed{0} \end{bmatrix}$$

The optimum situation is to assign courier firm A to delivery 3,
courier firm B to delivery 1,
courier firm C to delivery 2,
and courier firm D to delivery 4.

This would have a total cost of $140 + $220 + $90 + $120 = $570

Note: • Step 4, the drawing of the lines could be done mentally.

• If in step 4 the zeros are crossed through using less lines than there are rows (or columns) in the matrix, further adjustment of the matrix is necessary, as will be explained soon.

• The listing of all 24 possible allocations of courier firms to deliveries is shown below. The reader should confirm that the optimum solution given by the algorithm is the same as obtained from the list of all possibilities.

ABCD:	$250	+	$110	+	$160	+	$120	=	$640
ABDC:	$250	+	$110	+	$170	+	$130	=	$660
ACBD:	$250	+	$90	+	$150	+	$120	=	$610
ACDB:	$250	+	$90	+	$170	+	$110	=	$620
ADBC:	$250	+	$140	+	$150	+	$130	=	$670
ADCB:	$250	+	$140	+	$160	+	$110	=	$660
BACD:	$220	+	$100	+	$160	+	$120	=	$600
BADC:	$220	+	$100	+	$170	+	$130	=	$620
BCAD:	$220	+	$90	+	$140	+	$120	=	$570
BCDA:	$220	+	$90	+	$170	+	$120	=	$600
BDAC:	$220	+	$140	+	$140	+	$130	=	$630
BDCA:	$220	+	$140	+	$160	+	$120	=	$640
CABD:	$220	+	$100	+	$150	+	$120	=	$590
CADB:	$220	+	$100	+	$170	+	$110	=	$600
CBAD:	$220	+	$110	+	$140	+	$120	=	$590
CBDA:	$220	+	$110	+	$170	+	$120	=	$620
CDAB:	$220	+	$140	+	$140	+	$110	=	$610
CDBA:	$220	+	$140	+	$150	+	$120	=	$630
DABC:	$260	+	$100	+	$150	+	$130	=	$640
DACB:	$260	+	$100	+	$160	+	$110	=	$630
DBAC:	$260	+	$110	+	$140	+	$130	=	$640
DBCA:	$260	+	$110	+	$160	+	$120	=	$650
DCAB:	$260	+	$90	+	$140	+	$110	=	$600
DCBA:	$260	+	$90	+	$150	+	$120	=	$620

Question **2** of Exercise 8A involved the following information:

	Load 1	Load 2	Load 3
Firm A	$100	$140	$150
Firm B	$90	$170	$100
Firm C	$110	$140	$130

To minimise total cost the allocation was: firm A to do load 1,
firm B to do load 3,
firm C to do load 2.

The total cost would then be $340.

Follow the steps of the algorithm as shown below to see that it gives the same solution:

The cost matrix:

$$\begin{bmatrix} 100 & 140 & 150 \\ 90 & 170 & 100 \\ 110 & 140 & 130 \end{bmatrix}$$

Subtract the smallest number in each row from every number in that row.

$$\begin{bmatrix} 0 & 40 & 50 \\ 0 & 80 & 10 \\ 0 & 30 & 20 \end{bmatrix}$$

Subtract the smallest number in each column from every entry in that column.

$$\begin{bmatrix} 0 & 10 & 40 \\ 0 & 50 & 0 \\ 0 & 0 & 10 \end{bmatrix}$$

Draw as few horizontal and/or vertical straight lines as possible through all of the zeros. If the number of lines equals the number of rows (or columns) of the matrix then a solution can be found from this matrix.

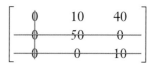

Determine the solution and relate identified locations in matrix back to the original.

$$\begin{bmatrix} \boxed{0} & 10 & 40 \\ 0 & 50 & \boxed{0} \\ 0 & \boxed{0} & 10 \end{bmatrix}$$

I.e., as before, firm A to do load 1 ($100), firm B to do load 3 ($100) and firm C to do load 2 ($140). Total cost $340.

ISBN 9780170395069

If drawing the horizontal and/or vertical lines on the matrix to cover the zeros can be done using fewer lines than there are rows (or columns) in the matrix we need to adjust the matrix so that a solution can be found.

For example, consider:

	1	2	3
A	8	5	17
B	11	7	18
C	9	8	15

The cost matrix:

$$\begin{bmatrix} 8 & 5 & 17 \\ 11 & 7 & 18 \\ 9 & 8 & 15 \end{bmatrix}$$

Subtract the smallest number in each row from every number in that row.

$$\begin{bmatrix} 3 & 0 & 12 \\ 4 & 0 & 11 \\ 1 & 0 & 7 \end{bmatrix}$$

Subtract the smallest number in each column from every entry in that column.

$$\begin{bmatrix} 2 & 0 & 5 \\ 3 & 0 & 4 \\ 0 & 0 & 0 \end{bmatrix}$$

Draw as few horizontal and/or vertical straight lines as possible through all of the zeros. The number of lines is less than the number of rows (or columns) the matrix has. Further adjustments needed.

$$\begin{bmatrix} 2 & 0 & 5 \\ 3 & 0 & 4 \\ 0 & 0 & 0 \end{bmatrix}$$

For the numbers that do not have a line drawn through them (2, 5, 3, 4), take the smallest of these numbers (2) from each of these numbers and also add this smallest number to any location where two of the straight lines intersect.

$$\begin{bmatrix} 0 & 0 & 3 \\ 1 & 0 & 2 \\ 0 & 2 & 0 \end{bmatrix}$$

The zeros now require three lines so a solution is possible (if not repeat the previous two steps). Determine the solution and relate the identified locations in matrix back to the original.

$$\begin{bmatrix} \boxed{0} & 0 & 3 \\ 1 & \boxed{0} & 2 \\ 0 & 2 & \boxed{0} \end{bmatrix}$$

A should be assigned to task 1. 'Cost' is 8 units.
B should be assigned to task 2. 'Cost' is 7 units.
C should be assigned to task 3. 'Cost' is 15 units.

Total 'cost' is 30 units.

The algorithm we are using here is called **The Hungarian Algorithm**, so named by Harold Kuhn in 1955, because the method was based on previous work by two Hungarian mathematicians, König and Egerváry.

Three examples of the algorithm are shown below. In each case the initial table shows the 'cost' of allocating each of A, B, C, … to tasks 1, 2, 3, … This 'cost' could be dollars, or hours, or distance travelled etc, and the task is to minimise the total cost. Check through each example to confirm that you too could obtain the solutions in this way.

	1	2	3	4
A	8	7	15	6
B	10	6	15	7
C	11	10	12	5
D	9	7	13	8

	1	2	3	4
A	60	50	120	70
B	50	60	100	80
C	55	80	140	80
D	80	80	110	70

	1	2	3
A	80	60	90
B	100	70	110
C	90	100	100

↓

$$\begin{bmatrix} 8 & 7 & 15 & 6 \\ 10 & 6 & 15 & 7 \\ 11 & 10 & 12 & 5 \\ 9 & 7 & 13 & 8 \end{bmatrix}$$

$$\begin{bmatrix} 60 & 50 & 120 & 70 \\ 50 & 60 & 100 & 80 \\ 55 & 80 & 140 & 80 \\ 80 & 80 & 110 & 70 \end{bmatrix}$$

$$\begin{bmatrix} 80 & 60 & 90 \\ 100 & 70 & 110 \\ 90 & 100 & 100 \end{bmatrix}$$

↓

$$\begin{bmatrix} 2 & 1 & 9 & 0 \\ 4 & 0 & 9 & 1 \\ 6 & 5 & 7 & 0 \\ 2 & 0 & 6 & 1 \end{bmatrix}$$

$$\begin{bmatrix} 10 & 0 & 70 & 20 \\ 0 & 10 & 50 & 30 \\ 0 & 25 & 85 & 25 \\ 10 & 10 & 40 & 0 \end{bmatrix}$$

$$\begin{bmatrix} 20 & 0 & 30 \\ 30 & 0 & 40 \\ 0 & 10 & 10 \end{bmatrix}$$

↓

$$\begin{bmatrix} 0 & 1 & 3 & 0 \\ 2 & 0 & 3 & 1 \\ 4 & 5 & 1 & 0 \\ 0 & 0 & 0 & 1 \end{bmatrix}$$

$$\begin{bmatrix} 10 & 0 & 30 & 20 \\ 0 & 10 & 10 & 30 \\ 0 & 25 & 45 & 25 \\ 10 & 10 & 0 & 0 \end{bmatrix}$$

$$\begin{bmatrix} 20 & 0 & 20 \\ 30 & 0 & 30 \\ 0 & 10 & 0 \end{bmatrix}$$

↓

$$\begin{bmatrix} \boxed{0} & 1 & 3 & 0 \\ 2 & \boxed{0} & 3 & 1 \\ 4 & 5 & 1 & \boxed{0} \\ 0 & 0 & \boxed{0} & 1 \end{bmatrix}$$

Assign A to 1. Cost = 8
Assign B to 2. Cost = 6
Assign C to 4. Cost = 5
Assign D to 3. Cost = 13

Total cost = 32

$$\begin{bmatrix} 10 & \boxed{0} & 20 & 10 \\ 0 & 10 & \boxed{0} & 20 \\ \boxed{0} & 25 & 35 & 15 \\ 20 & 20 & 0 & \boxed{0} \end{bmatrix}$$

Assign A to 2. Cost = 50
Assign B to 3. Cost = 100
Assign C to 1. Cost = 55
Assign D to 4. Cost = 70

Total cost = 275

$$\begin{bmatrix} 0 & 0 & 0 \\ 10 & \boxed{0} & 10 \\ 0 & 30 & 0 \end{bmatrix}$$

↓

Multiple solutions.
Assign B to 2
Cost = 70
Then either
A to 1 (cost 80) and
C to 3 (cost 100).
     Or
A to 3 (cost 90) and
C to 1 (cost 90).

Either way:
  Total cost = 250

Suppose we want a maximum

Question 6 of Exercise 8A involved assigning people to machines in such a way that the total number of components produced was **maximised**.

	Machine 1	Machine 2	Machine 3	Machine 4
Operator A	60	90	105	80
Operator B	80	90	110	85
Operator C	65	95	110	90
Operator D	75	85	115	75

To use the Hungarian algorithm for a situation requiring a **maximum** we first note the largest number in the table, 115 in the table above, subtract every number in the table from this largest number, **and then proceed as we did when determining a minimum.**

Subtract each entry in the table from 115 to obtain the matrix on the right.

$$\begin{bmatrix} 55 & 25 & 10 & 35 \\ 35 & 25 & 5 & 30 \\ 50 & 20 & 5 & 25 \\ 40 & 30 & 0 & 40 \end{bmatrix}$$

Subtract the smallest number in each row from every number in that row.

$$\begin{bmatrix} 45 & 15 & 0 & 25 \\ 30 & 20 & 0 & 25 \\ 45 & 15 & 0 & 20 \\ 40 & 30 & 0 & 40 \end{bmatrix}$$

Subtract the smallest number in each column from every entry in that column.

$$\begin{bmatrix} 15 & 0 & 0 & 5 \\ 0 & 5 & 0 & 5 \\ 15 & 0 & 0 & 0 \\ 10 & 15 & 0 & 20 \end{bmatrix}$$

Draw as few horizontal and/or vertical straight lines as possible through all of the zeros. If the number of lines equals the number of rows (or columns) of the matrix then a solution can be found from this matrix. If not, adjust as before.

$$\begin{bmatrix} 15 & 0 & 0 & 5 \\ 0 & 5 & 0 & 5 \\ 15 & 0 & 0 & 0 \\ 10 & 15 & 0 & 20 \end{bmatrix}$$

Determine the solution and relate identified locations in matrix back to original.

$$\begin{bmatrix} 15 & \boxed{0} & 0 & 5 \\ \boxed{0} & 5 & 0 & 5 \\ 15 & 0 & 0 & \boxed{0} \\ 10 & 15 & \boxed{0} & 20 \end{bmatrix}$$

I.e., assign operator A to machine 2, operator B to machine 1, operator C to machine 4 and operator D to machine 3.

Total number of components made is then 375 (= 90 + 80 + 90 + 115).

Check that this answer was the answer given for question 6 of exercise 8A.

8. Assignment problems ●●●●●●●●

Note: The Hungarian algorithm can be used when we are assigning one 'person per task' and one 'task per person'. It relies on the fact, not proven here, that adding the same number to every element in a row or column of the cost matrix (or subtracting the same number from every element in a row or column of the cost matrix) does not change the 'location' of the optimum arrangement. Once we have this location we can return to the original matrix to obtain the cost details.

Exercise 8B

For each of questions **1** to **4** use the Hungarian algorithm to allocate the companies A, B, C, ... to the tasks 1, 2, 3, ... such that each company is allocated a different task and the total 'cost' is **minimised**, where the units of 'cost' of each item being assigned to each task is as given by the numbers in the table or bipartite graph. As well as stating the assignments of company to task, also state the total cost.

1

	Task 1	Task 2
Company A	95	72
Company B	105	86

2

	Task 1	Task 2	Task 3
Company A	70	110	120
Company B	50	140	150
Company C	100	170	140

3

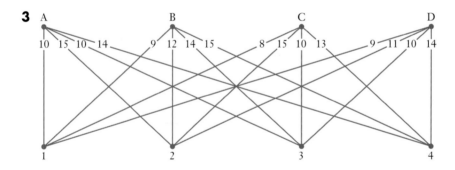

4

	Task 1	Task 2	Task 3	Task 4	Task 5	Task 6
Company A	20	30	20	30	20	30
Company B	24	20	12	35	10	28
Company C	30	25	20	55	21	38
Company D	25	44	21	36	20	30
Company E	20	32	20	40	22	35
Company F	40	50	20	48	25	40

ISBN 9780170395069

For each of questions **5** to **9** use the Hungarian algorithm to allocate the companies A, B, C, … to the tasks 1, 2, 3, … such that each company is allocated a different task and the total 'benefit' is **maximised**, where the units of 'benefit' of each company being assigned to each task is as given by the numbers in the table or bipartite graph. As well as stating the assignment of company to tasks also state the benefit of each assignment.

5

	Task 1	Task 2
Company A	250	120
Company B	240	90

6

	Task 1	Task 2	Task 3
Company A	29	29	31
Company B	26	24	28
Company C	22	33	22

7

	Task 1	Task 2	Task 3	Task 4
Company A	120	240	120	560
Company B	150	250	140	620
Company C	110	220	100	610
Company D	160	210	90	540

8

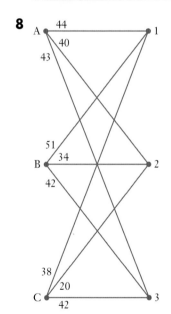

9

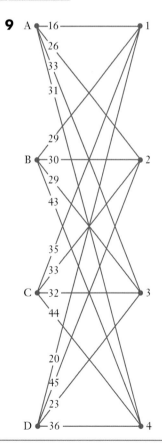

8. Assignment problems ●●●●●●●●

10 A company has four regional 'sales reps' with one situated in Adelaide, one in Brisbane, one in Hobart and one in Perth. The company plans to open offices in Broome, Darwin, Melbourne and Sydney and, at least initially, wants each of the existing sales reps to oversee the new reps in these new offices, with one existing rep per new office. This overseeing role is likely to involve a fair bit of travel so the company decides to attach each existing rep to the new office in a way that minimises the total cost of a return airfare for each existing rep to visit the new office they are overseeing.

The return airfares between the locations are as follows:

	Broome	Darwin	Melbourne	Sydney
Adelaide	$1080	$780	$240	$380
Brisbane	$1460	$570	$340	$270
Hobart	$1750	$1120	$290	$550
Perth	$720	$680	$440	$580

a Just by looking at the table try to write down what you think the optimal allocation of existing reps to new offices would be.

b Clearly showing your use of the Hungarian algorithm determine the optimal allocation.

11 a Allocate four delivery jobs to four courier companies with each company getting one of the jobs and total costs being kept to a minimum, and state this minimum, given that the cost of each job with each company is as shown below.

	Job 1	Job 2	Job 3	Job 4
Quick Courier Co.	$290	$70	$290	$150
Speedy Courier Co.	$320	$70	$260	$160
Deliverit Couriers	$460	$80	$320	$180
Competitive Couriers	$280	$60	$250	$140

b Suppose the job description for job 4 changed, causing every one of the four courier companies to increase their price for job 4 by $50, all other jobs remaining the same as in the above table. Would this alter the optimum allocation of company to job? If it would, state the new allocation.

c Suppose instead that it was job 2 that had its job description changed and this caused all four companies to multiply their price for that job by 4, all other jobs remaining as in the table. Would this alter the optimum allocation of company to job? If it would, state the new allocation.

d Suppose instead that just one of the companies is to be used for all four jobs. Which company would this be if again the aim is to minimise total cost?

12 A real estate company is allocating areas to each of its five salespeople, Jack, Jim, Judy, Nicci and Ti. Each salesperson is to have their own area, with the five areas labelled 1, 2, 3, 4 and 5.

An analysis of the areas, the selling history of the representatives and their existing profile in the areas suggest that the number of houses each representative should sell in a year, were they assigned to a particular area, would be as follows.

	Area 1	Area 2	Area 3	Area 4	Area 5
Jack	10	9	8	7	6
Jim	5	7	7	9	7
Judy	9	7	9	10	8
Nicci	8	6	10	10	10
Ti	11	12	8	13	10

Allocate the representatives to the areas in such a way that will maximise the total number of houses the five representatives should sell in a year.

What will this maximum total be?

13 A mining company employs four new engineers, Ashok, Daniel, Kirsten and Shani, and wishes to locate one at each of its four major mining locations. Travel and relocation expenses are available to all four and the company decides to send the engineers to the location that minimises the total travel and relocation costs. These costs are estimated to be as follows.

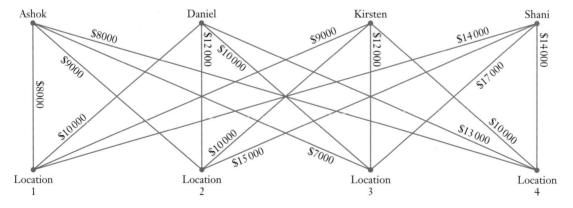

Allocate each engineer to the appropriate mining location.

14 A construction company has one mobile crane available from each of its five equipment storage bases. It needs to send these five mobile cranes to five job locations, with one crane per location. The distances involved for each crane to get to each location are as shown in the following table, in kilometres.

	Location 1	Location 2	Location 3	Location 4	Location 5
Crane A	40	90	30	70	60
Crane B	30	70	40	60	30
Crane C	40	30	70	40	30
Crane D	50	80	40	50	90
Crane E	50	60	40	40	70

Allocate each crane to a location such that the total distance travelled is minimised.

15 A car making company has five factories, A, B, C, D and E each producing all of the five major models of car that the company produces:

The Arcain   The Bijou   The Charger   The Devine   The Exeon

The company wishes to reorganise the factories and have one factory only producing the Arcain, another only producing the Bijou and so on.

Estimates suggest that the numbers each factory could produce of each model in a year, if they were dedicated to only producing the one model, would be as follows.

	Arcain	Bijou	Charger	Devine	Exeon
Factory A	230 000	250 000	200 000	190 000	180 000
Factory B	150 000	150 000	130 000	120 000	120 000
Factory C	320 000	330 000	300 000	280 000	290 000
Factory D	180 000	170 000	120 000	150 000	140 000
Factory E	240 000	200 000	200 000	180 000	180 000

The Bijou is the best-selling model so the company knows it wants to locate the construction of that model at its biggest factory, Factory C.

Allocate a factory to each model so each factory produces only one model, all five models are produced, the Bijou is at factory C, and the total number of models produced in a year is maximised.

ISBN 9780170395069

Suppose the number of people ≠ the number of tasks

In all of the assignment situations encountered so far in this chapter, the number of items we have been assigning (be they swimmers, trucking companies, machine operators, etc) has always been the same as the number of things we have assigned these items to (be they swimming legs in a relay, loads to be delivered, machines, etc.). Such assignment problems are sometimes referred to as being *balanced*. The cost matrix will be a square matrix. Suppose the number of items we are assigning does not match the number of items we are assigning them to? In such cases we assign a *dummy row*, or *dummy column*, as appropriate, with all of its entries equal to zero. In this way we **create a square matrix** and the Hungarian algorithm can be used. The next two examples illustrate this technique.

EXAMPLE 2

The director of a company decides that he wants a representative from the company at each of the three international trade conferences that are coming up. The director wishes to send a different manager to each, chosen from the four managers the company has based at its various branches, making the selection on the basis of the arrangement that keeps travel costs to a minimum. The associated travel costs are as follows:

	Conference 1	Conference 2	Conference 3
Manager A	$1040	$780	$2140
Manager B	$1150	$560	$1750
Manager C	$1780	$1250	$1350
Manager D	$1560	$790	$1215

What should the allocation of managers be?

Solution

Inserting a dummy column with all entries equal to zero gives the square matrix on the right.

$$\begin{bmatrix} 1040 & 780 & 2140 & 0 \\ 1150 & 560 & 1750 & 0 \\ 1780 & 1250 & 1350 & 0 \\ 1560 & 790 & 1215 & 0 \end{bmatrix}$$

With a zero in each row subtracting the lowest number in each row from every entry in that row will leave the matrix unchanged.

Subtract the smallest number in each column from every number in that column.

$$\begin{bmatrix} 0 & 220 & 925 & 0 \\ 110 & 0 & 535 & 0 \\ 740 & 690 & 135 & 0 \\ 520 & 230 & 0 & 0 \end{bmatrix}$$

Four horizontal and/or vertical lines are needed to cover all of the zeros so a solution can be found from this matrix.

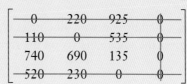

Determine the solution:

$$\begin{bmatrix} \boxed{0} & 220 & 925 & 0 \\ 110 & \boxed{0} & 535 & 0 \\ 740 & 690 & 135 & \boxed{0} \\ 520 & 230 & \boxed{0} & 0 \end{bmatrix}$$

Relate identified locations in the matrix back to original, remembering that the final column is for a non-existent conference.

$$\begin{array}{c} & 1 & 2 & 3 & \\ A \\ B \\ C \\ D \end{array} \begin{bmatrix} \boxed{1040} & 780 & 2140 & 0 \\ 1150 & \boxed{560} & 1750 & 0 \\ 1780 & 1250 & 1350 & \boxed{0} \\ 1560 & 790 & \boxed{1215} & 0 \end{bmatrix}$$

Hence send manager A to conference 1.　(for a cost of $1040)
Send manager B to conference 2.　(for a cost of $560)
Send manager D to conference 3.　(for a cost of $1215)
　　　　　　　　　　　　　　　　(Total cost: $2815)

Instead of making a dummy column with all entries equal to zero, try the previous example with the dummy column having all of its entries equal to the largest number in the matrix (i.e. 2140). Does that give the same solution?

EXAMPLE 3

An international company has three locations across the world, each with its own director. The three directors decide they should each make a fact-finding trip to one of four other countries to explore export opportunities. They choose four countries and decide that between them they will visit three of these four with each director visiting one country. The associated travel costs for each director to visit each country are as follows:

	Country 1	Country 2	Country 3	Country 4
Director A	$10100	$13300	$11500	$7200
Director B	$9300	$8200	$7900	$6300
Director C	$12700	$11700	$14200	$9500

Allocate each director to one of the countries so that each is attending a different country from the other two and total travel costs are kept to a minimum.

Solution

Inserting a dummy row with all entries equal to zero gives the square matrix on the right.

$$\begin{bmatrix} 10100 & 13300 & 11500 & 7200 \\ 9300 & 8200 & 7900 & 6300 \\ 12700 & 11700 & 14200 & 9500 \\ 0 & 0 & 0 & 0 \end{bmatrix}$$

Subtracting the smallest number in each row from every number in that row gives the matrix on the right.

With a zero in each column, subtracting the smallest number in each column will not change the matrix.

$$\begin{bmatrix} 2900 & 6100 & 4300 & 0 \\ 3000 & 1900 & 1600 & 0 \\ 3200 & 2200 & 4700 & 0 \\ 0 & 0 & 0 & 0 \end{bmatrix}$$

Drawing horizontal and vertical lines indicates further adjustment needed.

$$\begin{bmatrix} 2900 & 6100 & 4300 & 0 \\ 3000 & 1900 & 1600 & 0 \\ 3200 & 2200 & 4700 & 0 \\ 0 & 0 & 0 & 0 \end{bmatrix}$$

Adjust:

$$\begin{bmatrix} 1300 & 4500 & 2700 & 0 \\ 1400 & 300 & 0 & 0 \\ 1600 & 600 & 3100 & 0 \\ 0 & 0 & 0 & 1600 \end{bmatrix}$$

Drawing lines indicates yet further adjustment needed.

$$\begin{bmatrix} 1300 & 4500 & 2700 & 0 \\ 1400 & 300 & 0 & 0 \\ 1600 & 600 & 3100 & 0 \\ 0 & 0 & 0 & 1600 \end{bmatrix}$$

Adjust:

$$\begin{bmatrix} 700 & 3900 & 2100 & 0 \\ 1400 & 300 & 0 & 600 \\ 1000 & 0 & 2500 & 0 \\ 0 & 0 & 0 & 2200 \end{bmatrix}$$

Drawing lines indicates a solution can now be determined.

$$\begin{bmatrix} 700 & 3900 & 2100 & 0 \\ 1400 & 300 & 0 & 600 \\ 1000 & 0 & 2500 & 0 \\ 0 & 0 & 0 & 2200 \end{bmatrix}$$

Determine the solution and relate identified locations in matrix back to original, remembering that the bottom row is for a non-existent director.

$$\begin{bmatrix} 700 & 3900 & 2100 & \boxed{0} \\ 1400 & 300 & \boxed{0} & 600 \\ 1000 & \boxed{0} & 2500 & 0 \\ \boxed{0} & 0 & 0 & 2200 \end{bmatrix}$$

Director A should visit country 4.   (Cost: $7200)
Director B should visit country 3.   (Cost: $7900)
Director C should visit country 2.   (Cost: $11 700)
                    (Total cost: $26 800)

Exercise 8C

1 Three people, Alex, Ben and Theo, are each to be assigned a task from tasks 1, 2, 3 and 4, with each person having just one task, i.e. just three of the tasks will be assigned. The time each person takes to do each task, in minutes, is as shown below. The assignment of each person to a task is to be done in such a way that the total of the three times is to minimised.

Clearly showing your use of the Hungarian algorithm, assign each person to a task according to the above requirements.

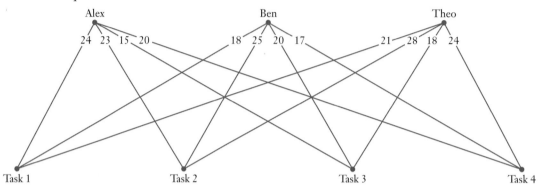

Under this minimal total time arrangement, if all three people start their assigned task at the same time, how many minutes later will all three have finished (assuming any early finishers do not assist others with their task)? Is there a different allocation arrangement that would reduce this 'time to finish three jobs when working simultaneously'?

2 An equipment hire company is asked to supply three industrial diggers, one to a worksite in Denton, one to a worksite in Abbysix and one to a worksite in Crampon.

The company has four such diggers available, one at the Horton storage facility, one at the Peel storage facility, one at the Washton storage facility and the fourth at the Wimsley storage facility. The distances, in kilometres, that each of these storage facilities is from each of the worksites is given in the following table:

	Denton	Abbysix	Crampon
Horton storage facility	27 km	25 km	25 km
Peel storage facility	58 km	22 km	18 km
Washton storage facility	34 km	36 km	51 km
Wimsley storage facility	31 km	55 km	54 km

Clearly showing your use of the Hungarian algorithm, assign three of the diggers, one to each worksite, such that the total distance to be travelled by the diggers to the worksites is minimised, and state this minimum distance.

3 A taxi company has four available taxis at various locations. The company receives three calls from separate individuals, Bill, Zarani and Tenielle, each requesting a taxi.

The estimated times each available taxi will take to get to each of the people requesting a taxi are as shown in the table.

	Bill	Zarani	Tenielle
Taxi A	9	7	8
Taxi B	11	7	10
Taxi C	10	6	8
Taxi D	8	8	9

Assign a taxi to each person such that the total time taken by the three taxis used is the minimum possible, and state the time each person has to wait.

4 Four people from A, B, C, D and E are to be assigned jobs 1, 2, 3 and 4 with a different person for each job and one of the five people 'missing out'.

The profit that will be made from each person doing each job is as shown in the table below.

	Job 1	Job 2	Job 3	Job 4
A	$75	$175	$150	$120
B	$90	$220	$150	$130
C	$120	$240	$170	$160
D	$100	$200	$160	$140
E	$90	$180	$150	$150

Supposing you are going to use the Hungarian algorithm to assign people to jobs according to the above requirements and in a way that maximises the profit to the company.

Which are you going to do first:

- put in the dummy column and then subtract each number from the largest number, or
- subtract each number from the largest number and then put in the dummy column, or
- doesn't it matter?

Investigate, and find the solution to the assignment problem too, stating which person gets which job and the maximum profit.

Challenge

Time-and-motion experts are looking at the assignment of ten workers to ten tasks, one worker per task. The experts first measure the time taken by each of the workers to complete each task. They find that the varied experience and skill levels of the workers mean that not all workers complete the same task in the same time. The various times taken by each worker on each task, in minutes, is as given below.

	Tasks									
	1	**2**	**3**	**4**	**5**	**6**	**7**	**8**	**9**	**10**
Worker A	17	21	9	28	15	19	33	45	15	14
Worker B	15	19	10	25	17	15	28	37	28	19
Worker C	22	22	13	21	17	22	47	46	17	17
Worker D	16	22	10	23	16	17	29	39	22	21
Worker E	18	20	9	29	21	18	32	40	23	14
Worker F	18	21	11	30	17	24	33	38	19	15
Worker G	17	23	12	22	16	18	45	40	24	21
Worker H	25	20	15	23	15	28	29	42	25	22
Worker I	16	20	10	25	14	17	32	55	20	18
Worker J	17	24	12	28	17	21	28	36	20	15

At present the workers are assigned as follows:

Worker A to task 1	(17 mins)	Worker B to task 2	(19 mins)
Worker C to task 3	(13 mins)	Worker D to task 4	(23 mins)
Worker E to task 5	(21 mins)	Worker F to task 6	(24 mins)
Worker G to task 7	(45 mins)	Worker H to task 8	(42 mins)
Worker I to task 9	(20 mins)	Worker J to task 10	(15 mins)

Summing these times gives a total of 239 minutes.

The experts want to find the allocation of worker to task that would minimise this total.

1. Solve the above allocation problem yourself using the Hungarian algorithm.

2. See if there is an online Hungarian algorithm solver and, if so, see if it gives the same solution to the above problem as you obtained.

ISBN 9780170395069

Miscellaneous exercise eight

This miscellaneous exercise may include questions involving the work of this chapter, the work of any previous chapters, and the ideas mentioned in the Preliminary work section at the beginning of the book.

1 Copy and complete the following table. (1 year = 12 months = 52 weeks = 365 days.)

Compounding frequency	Nominal annual interest rate (%)	Effective annual interest rate (%)
Annual	10%	
Six monthly	10%	
Quarterly	10%	
Monthly	10%	
Weekly	10%	
Daily	10%	

2 To the nearest dollar, how much should be invested in an annuity paying 10% annual interest, compounded annually, to provide a regular annual income of $65 000 for exactly 20 years?

Determine the answer by

a using a recursive formula with an initial 'guess' of $560 000, and

b using the financial capability of some calculators or computer programs.

3 Suppose instead that the annual interest rate in the previous question had been 7.5% rather than 10%. Initially make a guess at how much would need to be invested to provide the same annual income, i.e. $65 000 for exactly 20 years, and then use recursion to determine the correct answer, to the nearest dollar.

4 A loan for $8000 is taken out with compound interest charged at a rate of 8% per annum compounded monthly. If repayments of $250 are made at intervals of one month after the start of the loan how much is owed immediately after the fifteenth repayment has been made?

5 The times for tasks A, B, C etc., shown in the project network on the right are in hours.

(The letter I has been deliberately omitted.)

Determine the minimum completion time and critical path for the network.

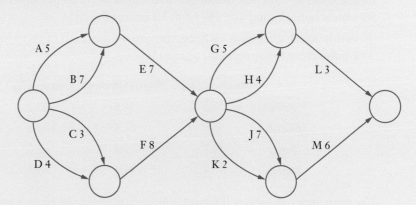

6 The network below shows the connections in a telecommunications system. The vertices represent the switching stations, capable of sending messages from one station to another. The numbers on the edges give the number of messages that can be sent from one station to another in an hour.

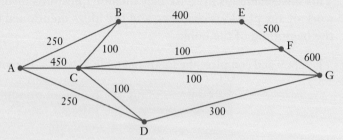

a Find the greatest number of messages that can go from A to G in an hour.

b Find the greatest number of messages that can go from A to D in an hour.

7 The network below shows a system of pathways that allow units to flow from a source A to a sink X. The units that can flow along each pathway are as indicated by the numbers next to each pathway.

Find the maximum number of units that can flow from A to X (and check your answer by finding a cut equal to your maximum flow).

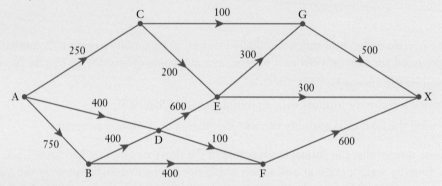

If the capacity of pathway EX were to be increased by 300 units, how would this alter the maximum flow from A to X?

8 Four swimmers, Julie, Micah, Fran and Marta are chosen to swim the medley relay for their school at the interschool swimming carnival. Each of the girls will swim one 'leg' of the relay. One will swim 50 metres backstroke, one will swim 50 metres breaststroke, one will swim 50 metres butterfly and one will swim 50 metres freestyle.

To decide which girl swims each leg, the coach times each of them over 50 metres of each style. The times recorded, in seconds, were as follows:

	Backstroke	Breaststroke	Butterfly	Freestyle
Julie	48.6	56.8	43.6	38.4
Micah	56.8	61.3	55.8	42.4
Fran	54.2	52.4	62.7	64.7
Marta	72.4	48.2	65.3	62.3

Use these times to suggest the most appropriate allocation of girls to strokes and suggest the total time for such a team.

ISBN 9780170395069

9 The following table gives the numbers of live cattle exported from Australia each year from 1988 to 2013. (Numbers are given to the nearest one hundred).

Year	1988	1989	1990	1991	1992	1993
Number of cattle	81 500	83 300	90 200	94 700	132 300	190 400

Year	1994	1995	1996	1997	1998	1999
Number of cattle	290 900	480 800	724 100	912 900	597 000	833 700

Year	2000	2001	2002	2003	2004	2005
Number of cattle	887 000	797 900	955 100	689 400	570 800	531 200

Year	2006	2007	2008	2009	2010	2011
Number of cattle	609 200	668 900	815 600	921 300	821 600	621 900

Year	2012	2013
Number of cattle	516 500	726 000

Source of data: Australian Bureau of Statistics.

With the assistance of a computer spreadsheet draw both the raw data and the five-point moving averages on the same time series line graph.

Write some sentences describing any trends in the numbers of live cattle exported from Australia per year over the period of time from 1988 to 2013.

10 Five police vehicles, A, B, C, D and E have to be sent to five locations, 1, 2, 3, 4 and 5, with one police vehicle per location. The estimated time, in minutes, for each police vehicle to reach each location, from where they are now, is given in the following bipartite graph.

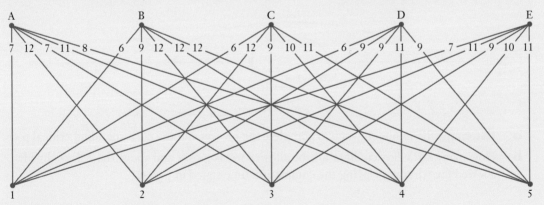

Allocate the police vehicles to the locations in such a way that the sum of the five travel times is minimised, and state the time taken for each vehicle to reach its allocated location.

---

11 A company expects the annual sales of one of their products next year to be 27 400 and anticipate that the only significant changes in sales from one quarter to another to be those due to anticipated seasonal effects.

   a If there is no quarterly seasonal effect on sales of this product roughly how many of the product should the company expect to sell each quarter?

   b In fact the sales are seasonal in nature with the seasonal indices for each of the first three quarters being as follows:

        1st Quarter: 0.72   2nd Quarter: 0.88   3rd Quarter: 1.06

      Predict the sales numbers for this product for each quarter of next year.

12 Calculate the total *length* of the minimum spanning tree for the network shown below and show the minimum spanning tree on a diagram. (The numbers on each edge gives the *length* of that edge in units.)

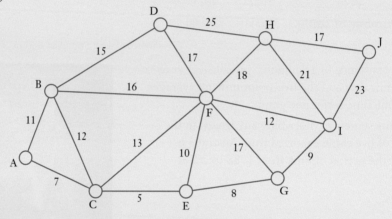

13 A company identifies a number of tasks involved in the completion of a particular project. The project network below shows these tasks labelled A to J with the numbers on each arc indicating the number of days each task requires.

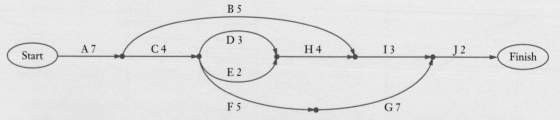

   a Determine the minimum time to complete the project and the corresponding critical path.

   b How many extra days could task E be allowed to take, over and above the 2 days already allowed for, without delaying the completion time stated in part **a**?

14 The table below shows the number of gas units used by a company every two months for a period of three years.

n	Year	Two-month period	Units used
1	1	1st two months of year	1238
2		2nd two months of year	1812
3		3rd two months of year	5084
4		4th two months of year	6320
5		5th two months of year	3778
6		6th two months of year	1052
7	2	1st two months of year	1382
8		2nd two months of year	2196
9		3rd two months of year	6416
10		4th two months of year	7148
11		5th two months of year	4870
12		6th two months of year	1634
13	3	1st two months of year	1502
14		2nd two months of year	2586
15		3rd two months of year	7460
16		4th two months of year	8216
17		5th two months of year	5122
18		6th two months of year	1790

a Determine the seasonal indices for each two-month period of the year using the average (mean) percentage method. Give answers in decimal form and correct to four decimal places.

b Use the seasonal indices to deseasonalise the number of units used for each two-month period. Give each deseasonalised figure to the nearest ten units.

c View the deseasonalised figures plotted against n to check that linear regression would be an appropriate model to use for this data pair.

d Use n and the deseasonalised data, D, to determine the least squares regression line $D = an + b$. (Give a correct to two decimal places and b to the nearest ten.)

e Use your regression line, and the seasonal indices to reseasonalise the data, to predict the number of units used for **i** the 1st two months of year 4,
ii the 4th two months of year 4.

f Suggest some events that could occur that would lead us to expect that the actual values for year 4 could differ markedly from our predicted values.

15 A company is planning the construction of a new shopping complex.

The company identifies a number of tasks, the time each task will take and the order in which they must be completed:

Task	Time	Order
A	1 week	• Tasks B, C and D can commence provided task A has finished.
B	3 weeks	• Task E can commence provided task B has finished.
C	1 week	• Task F can commence provided task C has finished.
D	2 weeks	• Task H can commence provided task D has finished.
E	3 weeks	• Tasks K and I can commence provided task H has finished.
F	4 weeks	• Tasks G and M can commence provided both E and F have finished.
G	6 weeks	
H	5 weeks	• Task J can commence provided both G and I have finished.
I	7 weeks	• Task L can commence provided both J and K have finished.
J	4 weeks	
K	9 weeks	
L	2 weeks	
M	12 weeks	

a Determine the minimum completion time and the critical path.

b What is the maximum time that task K could take without delaying the completion of the project?

c If the time for task I could be reduced to 4 weeks what would be the minimum completion time for the project?

16 A minimum spanning tree is needed for the network shown below. The number on each edge is the cost, in $1000s, of making the connection. Find the minimum spanning tree for the network and state its total cost.

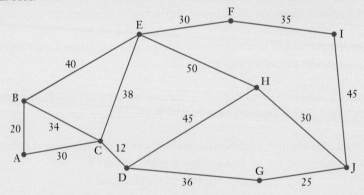

Before work is started on constructing the minimum spanning tree, it is found that a direct connection could be made from H to I. What would the cost of this connection need to be for it to be worthwhile to include in a new minimum spanning tree?

ISBN 9780170395069

17 One particular day, three machines are to be assigned to manufacturing different products, with each machine producing a different product. There are four products required but with just the three machines working one product will not be manufactured that day. The assignment of machine to product is to be done so that the total output, in total number of units produced for the day, is maximised.

Clearly showing your use of the Hungarian algorithm, allocate machines to products to achieve this, given that the numbers of units of each product that could be produced on each machine in the day is as follows:

	Product 1	Product 2	Product 3	Product 4
Machine A	2700	2300	3300	3300
Machine B	4200	3600	4100	4100
Machine C	4000	3500	3600	3300

18 A port authority wishes to build a railway network linking seven locations in the dockside complex. The network is to be constructed so that it will be possible to travel by rail from any one of the seven locations to any of the other six, perhaps not directly, but at least by going via other locations in the network. The locations are Goods In, Goods Out, Waiting, Customs, Dock 1, Dock 2 and Dock 3. The distances involved for the feasible rail connections from any of these locations to any of the others are given in metres in the table below.

	Goods In	Goods Out	Waiting	Customs	Dock 1	Dock 2	Dock 3
Goods In	–	–	420	–	300	–	–
Goods Out	–	–	–	150	–	–	210
Waiting	420	–	–	120	240	–	–
Customs	–	150	120	–	–	210	180
Dock 1	300	–	240	–	–	195	–
Dock 2	–	–	–	210	195	–	225
Dock 3	–	210	–	180	–	225	–

a Based on these distances determine the minimum length of track required for this network.

b In this 'minimum length network', which six pairs of locations would have a direct rail link? (Direct rail link means able to journey from one location to the other without having to pass through any of the other locations on the way.)

8. Assignment problems ●●●●●●●●

19 Five transport companies are asked to quote their price for four jobs involving the transportation of wide loads. The company asking for the quotes is likely to have several more such jobs in the future and wants to use a different company for each of the four jobs in order to see which company seems to give the best service. The five transport companies are *DG Transport*, *Haulage and Co*, *JA Movers*, *Move it and Shift it* and *Relocate Plus*. The prices each company quotes for the four jobs are as in the following table:

	Job 1	Job 2	Job 3	Job 4
DG Transport	$18 000	$14 600	$28 000	$16 500
Haulage and Co	$19 800	$12 800	$32 000	$18 000
JA Movers	$18 000	$13 500	$27 000	$17 500
Move it and Shift it	$24 000	$12 000	$25 000	$16 000
Relocate Plus	$22 000	$13 000	$26 500	$17 000

Allocate a transport company to each job with a different company for each job, one company missing out, and minimising the total cost of the four jobs according to the quotes given above. Also state what this minimum cost will be.

ISBN 9780170395069

ANSWERS

1 a No, the increase cannot be attributed to the natural increase in the population. Numbers are given *per 100 persons of working age*, thus the increase cannot simply be attributed to increase in population.

b

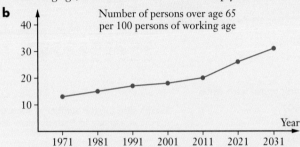

Number of persons over age 65
per 100 persons of working age

c The number of persons over age 65 per 100 persons of working age shows an increasing trend with time. The increase is steady at approximately 2 more people per 100, for each ten years from 1971 to 2011, and then increases to about 5 more people per 100, for each ten years from 2011 to 2031.

2 a

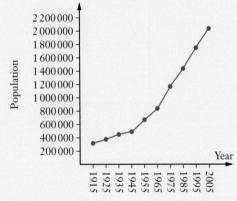

Line graph showing the population
of Western Australia, 1915 to 2005

b The population shows an increasing trend over the years from a population of just over 300 000 in 1915 to just over 2 million in 2005.

Earlier 10 year intervals in this time show an increase in population of about 60 000 people per ten years. Later 10 year intervals show an increase in population of about 300 000 people per ten years.

c The figures suggest that the population of Western Australia in 1950 was approximately 580 000. This prediction involves interpolating between two known points (1945, 490 000) and (1955, 669 000). Hence the prediction should be very reliable.

d Continuing the trend suggests a population in 2025 of about 3 million (see graph on the right which shows two possible continuations to the trend). However extrapolation is involved as we are going beyond the known data points to make our prediction. Continuing the line as shown is an attempt to predict what will happen, but a lot could change between the 2005 figure and 2025 so with this degree of extrapolation involved the prediction should not be viewed as being particularly reliable. If we had to predict it would perhaps be safer to predict in a range, say 2.8 million to 3.2 million.

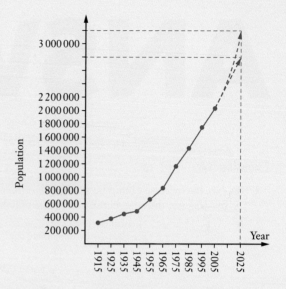

3 a The number of short term visitors shows an increasing trend with time.

The increase seems reasonably steady from 1992–1997, then the rate of increase slows, with numbers actually declining in 2003 before picking up again. From 2005 to 2010 numbers of visitors are reasonably steady before increasing again from 2011 to 2013.

Graph showing the number of short term visitors arriving in Australia each year

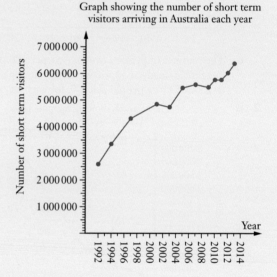

b The graph suggest visitor numbers of approximately 4 600 000 for 1999.

This prediction involves interpolation so it should be reliable.

The flattening of the graph around 2005 to 2010 makes extrapolation to 2020 difficult to do with any confidence. The 2013 figure perhaps suggests the upward trend continues so a prediction of approximately 7 500 000 for 2020 might be reasonable.

We are not extrapolating that far beyond the known values but extrapolation is involved so our prediction could not be relied upon with any great confidence.

ISBN 9780170395069

4

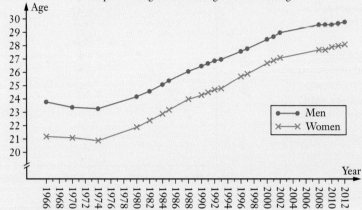

Graph showing the median age at first marriage

Written response not included here.

Compare your response with those of others in your class and discuss with your teacher.

5 $N = 494Y - 917\,000$

6 a Number of properties managed $= 290 + 11.4t$

b 655

7 a $P = 0.8x + 17.2$

b

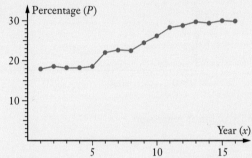

Over the sixteen years the percentages of births delivered by caesarean section show an increasing trend with most of the increase being from year 5 to year 11.

For the first five years the percentage remains level at around 18%.

For the last four years the percentage remains level at around 30%.

8 a

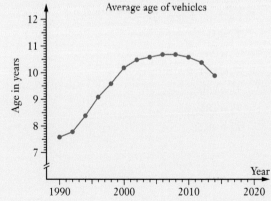

Average age of vehicles

b The data is not suitable for linear regression. The graph shows that the distribution of points does not follow a linear pattern, nor anything that could be well approximated by a linear pattern.

Predicted values given by the 'best straight line' as determined using linear regression would overestimate ages for the early and late years, and underestimate ages for the middle years – see diagram on the right in which the least squares linear regression line is shown as a broken line.

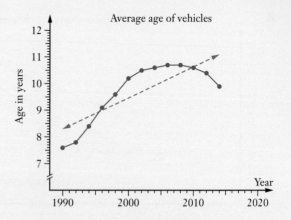

Average age of vehicles

c Predicted average age of vehicle for 1999 would be about 9.9 years. We have data for 1998 and 2000 so our value for 1999 involves interpolation. It should therefore be a prediction that could be relied on with confidence.

We cannot predict with any great confidence what the average age of vehicles will be in 2020 as 2020 is well beyond the years plotted (i.e. extrapolation is involved). If decline in later years were to continue, the figure for 2020 might perhaps be around 7.8. No great confidence in reliability of this prediction though.

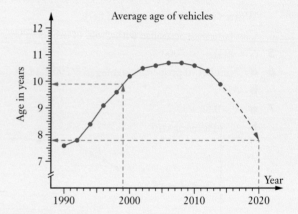

Average age of vehicles

9 Viewing the graph of Y plotted against t confirms the unsuitability of linear regression. $Y = 4.69 \times t^{2.145}$.

Miscellaneous exercise one   PAGE 16

1 $2316.50

2 $3276

3 A: $y = x$ B: $y = 2x + 1$ C: $y = 0.5x + 1$ D: $x = 3$ E: $y = 2$ F: $y = -2x - 3$

4 a $T_1 = 25,$ $T_2 = 27,$ $T_3 = 29,$ $T_4 = 31,$ $T_5 = 33$

b $T_1 = 32,$ $T_2 = 30.5,$ $T_3 = 29,$ $T_4 = 27.5,$ $T_5 = 26$

c $T_1 = 64,$ $T_2 = 96,$ $T_3 = 144,$ $T_4 = 216,$ $T_5 = 324$

d $T_1 = 4000,$ $T_2 = 2000,$ $T_3 = 1000,$ $T_4 = 500,$ $T_5 = 250$

e $T_1 = 5,$ $T_2 = 17,$ $T_3 = 41,$ $T_4 = 89,$ $T_5 = 185$

f $T_1 = 16,$ $T_2 = 20,$ $T_3 = 26,$ $T_4 = 35,$ $T_5 = 48.5$

5

Percentage of WA population living in Perth

Year	1911	1921	1933	1947	1954	1961	1966	1971	1976	1981	1986	1991
%	41%	51%	52%	60%	62%	65%	67%	68%	71%	71%	72%	73%

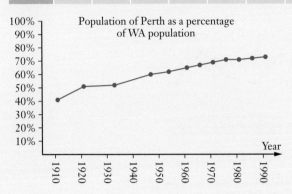

Perth population, as % of WA population, shows an increasing trend.

This Increase 'levels off' in later years but the percentage is still increasing.

b i For 1940 approx. 56%.

  1940 is between years for which the population is known so interpolation is involved.

  Estimate should be reliable.

ii For 2020 approx. 80%.

  Extrapolation is involved and 2020 is a long way beyond known values. Estimate may not be too reliable.

  (Note: Linear regression gives 86% but not a particularly suitable model.)

Exercise 2A PAGE 23

1 3-point **2** 4-point **3** 4-point **4** 3-point **5** 5-point **6** 3-point

7 a and **b** See graph:

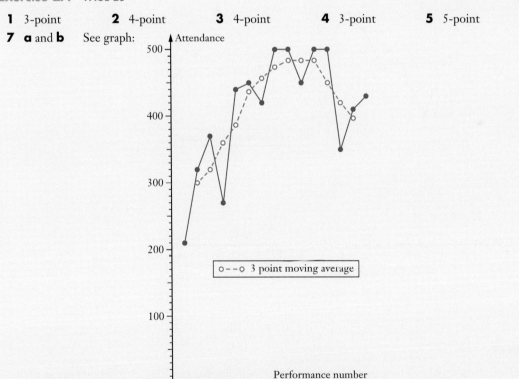

c Many factors such as cost, venue and actor availability, would need to be considered before we could decide if a sixth week would be 'a good idea'. However, attendance figures suggest a 6th week could possibly have given attendance similar to week 1 so on that basis alone, and if the attendance of week 1 was considered acceptable, a 6th week could have been a good idea.

ISBN 9780170395069

8 a Four-point moving averages are:

26.5, 26.25, 24.5, 24.5, 24, 23.5, 23.5, 23, 22.25, 22, 20.75.

b

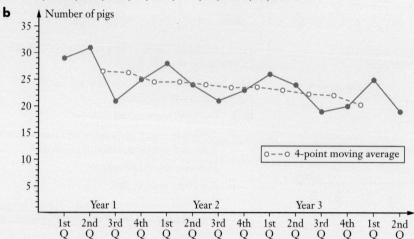

c Decreasing.

9 a A = 263,   B = 270,   C = 276,   D = 277,   E = 256.

b

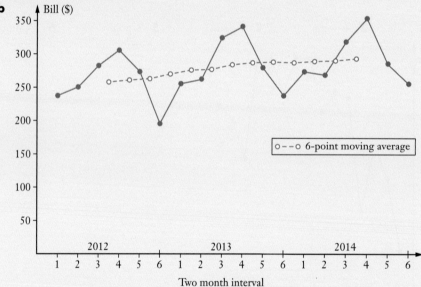

c Increasing.

Exercise 2B PAGE 29

1 Consumption of soft drink could well be seasonal in nature. Hence, with the data involving consumption per quarter year a 4-point average would be most suitable.

a A = 6940,   B = 6892,   C = 6788,   D = 6763,   E = 5511,   F = 6638,   G = 6662

2 a A = 176.5,   B = 180,   C = 179,   D = 182,   E = 183,   F = 184,   G = 181,
   H = 190,   I = 189,   J = 191.5

3

	Year 1				Year 2		
Summer	**Autumn**	**Winter**	**Spring**	**Summer**	**Autumn**	**Winter**	**Spring**
1228	364	640	1220	1436	276	752	1132

4-point MAs	863	915	893	921	899
4-point CMAs		889	904	907	910

4 a We would expect the production of fruit to be dependent on the weather and hence follow a seasonal pattern. Hence determining a seasonal moving average makes sense and with data collected on a monthly basis a 12-point moving average is sensible.

b

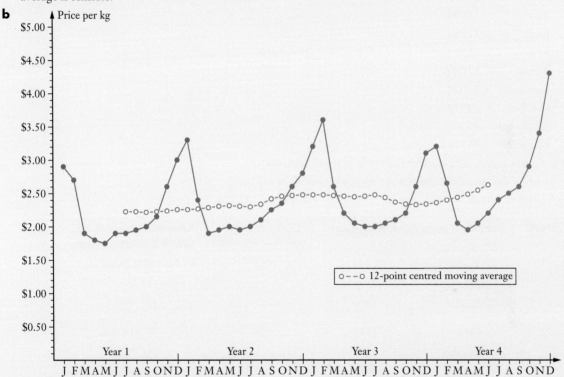

c Underlying trend shows a slight increase over time.

5 a The 5-point moving average would smooth the data more than the 3-point moving average.

b

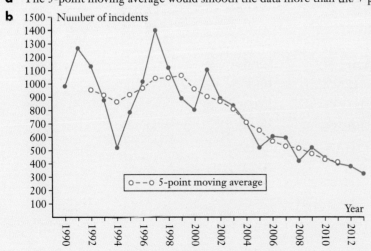

c In the years under consideration there is an initial downward trend followed by an upward trend, but then from about the year 2000 the trend has been decreasing.

Hence the expectation mentioned at the beginning of the question, i.e. that with increasing population we would expect the number of criminal incidents to increase, was not the case for this category of crime.

(From 1990 to 2013 the NSW population increased from 5.8 million to 7.4 million.)

Exercise 2C PAGE 36

1 The seasonal index for the fourth quarter is 98%.

2 The seasonal index for Wednesday is 94%.

3 a With no seasonal effect on sales the company should expect to sell roughly 13 000 each month.

   b Predicted sales for each month would be as follows:

January	11 440
February	10 660
March	9880
April	9620
May	10 790
June	12 740
July	13 520
August	13 650
September	14 040
October	14 430
November	16 380
December	18 850

4 The figures suggest that sales for the whole year will be 62 000.

For each of the remaining quarters the figures suggest sales will be:

2nd quarter   16 120,   3rd quarter   15 190,   4th quarter   13 330.

5

Week	Period of time	Attendance	Daily mean for the week	Attendance as % of week's daily mean
1	Friday	2346	2766	84.82%
	Saturday	3143		113.63%
	Sunday	2809		101.55%
2	Friday	2572	2929	87.81%
	Saturday	3258		111.23%
	Sunday	2957		100.96%
3	Friday	2987	3231	92.45%
	Saturday	3500		108.33%
	Sunday	3206		99.23%

	Friday	Saturday	Sunday
1st week	84.82%	113.63%	101.55%
2nd week	87.81%	111.23%	100.96%
3rd week	92.45%	108.33%	99.23%
Index (1 dp)	88.4%	111.1%	100.6%

ISBN 9780170395069

6	Year	Period of time	Units sold	Quarterly mean for the year	Units as % of year's quarterly mean
	1	1st quarter	54 000	52 650	102.56%
		2nd quarter	63 200		120.04%
		3rd quarter	45 200		85.85%
		4th quarter	48 200		91.55%
	2	1st quarter	48 500	47 250	102.65%
		2nd quarter	56 700		120.00%
		3rd quarter	41 300		87.41%
		4th quarter	42 500		89.95%
	3	1st quarter	51 800	49 675	104.28%
		2nd quarter	62 600		126.02%
		3rd quarter	40 500		81.53%
		4th quarter	43 800		88.17%

	1st quarter	2nd quarter	3rd quarter	4th quarter
1st year	102.56%	120.04%	85.85%	91.55%
2nd year	102.65%	120.00%	87.41%	89.95%
3rd year	104.28%	126.02%	81.53%	88.17%
Seasonal index (1 dp)	103.2%	122.0%	84.9%	89.9%

7	n	Week	Day	No. of calls	5-pt MA	Calls as % of daily average for week
	1		Mon	258	–	107.59%
	2		Tue	231	–	96.33%
	3	One	Wed	215	239.8	89.66%
	4		Thur	248	235.6	103.42%
	5		Fri	247	231.0	103.00%
	6		Mon	237	224.8	109.32%
	7		Tue	208	221.6	95.94%
	8	Two	Wed	184	216.8	84.87%
	9		Thur	232	215.8	107.01%
	10		Fri	223	213.0	102.86%
	11		Mon	232	212.8	113.06%
	12		Tue	194	208.2	94.54%
	13	Three	Wed	183	205.2	89.18%
	14		Thur	209	–	101.85%
	15		Fri	208	–	101.36%

	Week 1	Week 2	Week 3	Seasonal index
Monday	107.59%	109.32%	113.06%	109.99%
Tuesday	96.33%	95.94%	94.54%	95.60%
Wednesday	89.66%	84.87%	89.18%	87.90%
Thursday	103.42%	107.01%	101.85%	104.09%
Friday	103.00%	102.86%	101.36%	102.41%

Exercise 2D PAGE 43

(You may occasionally find that an answer you obtain varies slightly from the answer given here depending on when, and to what degree, rounding occurred.)

1 $22 300 (= $25 400 ÷ 1.14 and then given to the nearest $100).

2 173 700 units (= 132 000 ÷ 0.76 and then given to nearest 100).

3 a 69 530 people.   **b** 69 940 people.

4 The analysis predicts that 14 750 units will be sold in January.

5 The figures suggest that the actual weight of the fruit this summer will be 23 680 kg.

6 a Seasonal index for Jan–Apr: 107.25%
        Seasonal index for May–Aug: 75.56%
        Seasonal index for Sept–Dec: 117.20%

b

	Jan–Apr	May–Aug	Sept–Dec
Year 1	1.71	1.71	1.69
Year 2	1.74	1.87	1.76
Year 3	1.80	1.69	1.84
Year 4	2.03	2.01	1.99

7 a Seasonal index for Friday: 0.6436   Seasonal index for Saturday: 1.0540
        Seasonal index for Sunday: 1.1555   Seasonal index for Monday: 1.1469

	Friday	Saturday	Sunday	Monday
Year 1	14 290	14 080	14 070	14 690
Year 2	15 240	14 980	14 950	15 520
Year 3	14 680	14 440	14 890	14 660
Year 4	15 290	16 030	15 610	14 600

8 a

	Monday	Tuesday	Wednesday	Thursday	Friday	Saturday	Sunday
Week 1	$652	$643	$690	$688	$644	$645	$721
Week 2	$734	$760	$760	$787	$746	$707	$725
Week 3	$855	$837	$779	$751	$852	$895	$778

b Graph of deseasonalised data, D, plotted against t, as shown on the right, shows a reasonable linear trend. Hence linear regression is appropriate.

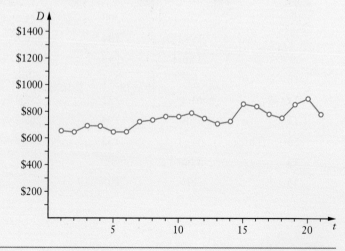

c $D = 9.64t + 639.1$

d For Monday of week 4 ($t = 22$) the regression line predicts value for D of \$851.2.
Hence prediction of real takings will be \$851.2 × 0.8242 = \$702, to the nearest dollar.

For Wednesday of week 4 ($t = 24$) the regression line predicts value for D of \$870.5.
Hence prediction of real takings will be \$870.5 × 0.5394 = \$470, to the nearest dollar.

For Saturday of week 4 ($t = 27$) the regression line predicts value for D of \$899.4.
Hence prediction of real takings will be \$899.4 × 1.4194 = \$1277, to the nearest dollar.

9 a Viewing the graph, not shown here, confirms reasonableness of using linear regression for (n, M) data.

b $M = 1.0994n + 26.031$

c

	Monday	Tuesday	Wednesday	Thursday	Friday
Week one	1.15132	0.82237	0.78947	0.92105	1.31579
Week two	1.07558	0.87209	0.78488	0.87209	1.39535
Week three	1.14428	0.79602	0.82090	0.87065	1.36816
Week four	1.04348	0.91304	0.86957	0.95652	1.21739
Mean (4 dp)	1.1037	0.8509	0.8162	0.9051	1.3242

d Missing figures are:

	Tuesday	Wednesday	Thursday
Week one	29	29	31
Week two	35	33	33
Week three	38	40	39
Week four	49	49	49

e Predictions for week 5:   Mon 54   Tues 43   Wed 42   Thur 47   Fri 71

f If week five involved things like 'pupil free days', public holidays, unusually severe weather making travel to school difficult, a flu epidemic in the area, school closure due to fire etc., our week five attendance predictions would not be reliable. (Also, we cannot assume that the trends of the first four weeks will necessarily continue even without these unusual events.)

10 a

	Rescues as a percentage of annual mean (As percentages and correct to 2 decimal places)		
	Oct/Nov	**Dec/Jan**	**Feb/Mar**
Season 1	56.10%	135.37%	108.54%
Season 2	47.37%	136.84%	115.79%
Season 3	51.43%	152.86%	95.71%
Season 4	53.13%	142.19%	104.69%
Seasonal indices (Nearest percent)	52%	142%	106%

b

	Oct/Nov	Dec/Jan	Feb/Mar
Season 1	88	78	84
Season 2	69	73	83
Season 3	69	75	63
Season 4	65	64	63

ISBN 9780170395069

c Viewing the graph, not shown here, confirms reasonableness of using linear regression for (n, D) data.

Using integer values of D: $D = 85.83 - 2n$.

(Using more accurate values of D: $D = 86.05 - 2.003n$.)

d

n	Season	Months	Predicted number of rescues
13		Oct/Nov	31
14	5	Dec/Jan	82
15		Feb/Mar	59

Miscellaneous exercise two   PAGE 48

1 $247.50

2 1

3 A seasonal index of 0.87 for February means that whatever it is that we are measuring for February tends to be 13% below the monthly average.

4

Season	Spring	Summer	Autumn	Winter
Deseasonalised data	1000	1552	1600	1875

5 The starting amount is $1 500 000.

6 The scattergraph indicates that in general the more recently manufactured the vehicle the higher the asking price. Given the number of points involved, and their distribution, the relationship between year of manufacture and asking price could be well represented by a linear model.

$p = 3561t - 697$

Likely price for a 2007 model: $31 350.

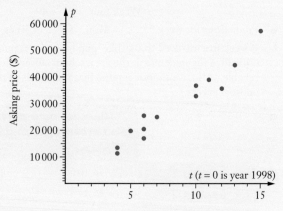

7 a The given revenues have a total of $3 760 000 and thus a quarterly average of $940 000. The first quarter is 20% down on this quarterly average (compared to the 15% down the seasonal index suggests), the second quarter is 21% up (compared to the expected 25%), the third quarter is 25% up (compared to the expected 12%) and the fourth quarter is 26% down (compared to the expected 22%). Thus the third quarter revenue is a surprise when compared to what we might expect from the seasonal indices.

b If all other factors remained the same then the third quarter increase in the revenue share above expectation could be as a result of the advertising campaign being successful.

8 a

t	Year	Quarter	Number sold in quarter	Number sold as percentage of quarterly mean for the year	Number sold seasonally adjusted
1	One	1st	1637	106.99%	1514
2		2nd	1489	97.32%	1537
3		3rd	1244	81.31%	1598
4		4th	1750	114.38%	1494
5	Two	1st	1405	108.16%	1299
6		2nd	1241	95.54%	1281
7		3rd	1012	77.91%	1300
8		4th	1538	118.40%	1313
9	Three	1st	1253	109.24%	1159
10		2nd	1121	97.73%	1157
11		3rd	852	74.28%	1095
12		4th	1362	118.74%	1162

	Year 1	Year 2	Year 3	Seasonal index
1st Quarter	106.99%	108.16%	109.24%	108.13%
2nd Quarter	97.32%	95.54%	97.73%	96.86%
3rd Quarter	81.31%	77.91%	74.28%	77.83%
4th Quarter	114.38%	118.40%	118.74%	117.17%

c Using integer values for seasonally adjusted figures: $S = -43.885t + 1611$
(Using more accurate values for seasonally adjusted figures: $S = -43.887t + 1611.1$)

d Predicted numbers sold for year 4 are:
1st quarter 1125, 2nd quarter 965, 3rd quarter 742, 4th quarter 1065.

Exercise 3A   PAGE 56

1 $2720   **2** 5 years   **3** 12.5%   **4** $2400   **5** 7.4%   **6** $7700

7 $6347.50   to the nearest cent.   **8** $9609.32   to the nearest cent.

9 Annual compounding gives interest of $13 123.85, to the nearest cent.
Monthly compounding would give $400.61 more interest, to the nearest cent.

10 Annual compounding gives interest of $463 570.00, to the nearest cent.
Daily compounding would give $26 568.34 more interest, to the nearest cent.

11 $6 965.59

12 It would take approximately 11.64 years for the investment to double in value. I.e. approximately 11 years and 8 months.

13 It would take almost 20 compoundings (19.89), i.e. just under 10 years (9.945 years).

	1st term	2nd term	3rd term	4th term	5th term	15th term
14	5,	8,	11,	14,	17.	47.
15	5,	2,	−1,	−4,	−7.	−37.
16	−10,	−5,	0,	5,	10.	60.
17	12.5,	15,	17.5,	20,	22.5.	47.5.
18	0.25,	0.5,	1,	2,	4.	4096.
19	20480,	30720,	46080,	69120,	103680.	5978711.25.
20	2621440,	1310720,	655360,	327680,	163840.	160.
21	2^{30},	2^{29},	2^{28},	2^{27},	2^{26}.	2^{16}.
22	5,	13,	29,	61,	125.	131069.
23	$1000,	$1300,	$1660,	$2092,	$2610.40.	$18758.78.
24	$5000,	$5000,	$5000,	$5000,	$5000.	$5000.
25	12,	23,	34,	45,	56.	166.
26	819200,	819200,	819200,	819200,	819200.	819200.
27	1,	1,	2,	3,	5.	610.
28	2,	3,	8,	19,	46.	309268.
29	5,	−5,	−5,	5,	5.	−5.
30	5,	−3,	−1,	3,	5.	−1.

Exercise 3B PAGE 59

Percentages given rounded to three decimal places where rounding is necessary.

1

Compounding frequency	Nominal annual interest rate (%)	Effective annual interest rate (%)
Annual	4%	4%
Six monthly	4%	4.04%
Quarterly	4%	4.060%
Monthly	4%	4.074%
Weekly	4%	4.079%
Daily	4%	4.081%

2

Compounding frequency	Nominal annual interest rate (%)	Effective annual interest rate (%)
Annual	8%	8%
Six monthly	8%	8.16%
Quarterly	8%	8.243%
Monthly	8%	8.300%
Weekly	8%	8.322%
Daily	8%	8.328%

3 The effective annual interest rates for 8% are not simply double the rates for 4%.

Doubling the value of 'i' in the formula $\left(1+\dfrac{i}{n}\right)^n - 1$ will not simply double the answer.

Exercise 3C   PAGE 63

1 $T_{n+1} = 1.08 \times T_n + \$200$ $\$11\,533.01$ is in the account at the end of ten years.

2 a $T_{n+1} = 1.005 \times T_n + \$100$ **b** Final value of account is $\$9817.01$. $(= T_{36} - \$100)$

3 $\$22\,474.58$ (nearest cent).   **4** $\$4780.04$ (nearest cent).   **5** $\$14\,773.77$ (nearest cent).

6 a 10.164% (correct to three decimal places).   **b** 7.590% (correct to three decimal places).

Exercise 3D   PAGE 65

1 a End of first year $\$195\,000$ **b** End of first year $\$187\,000$
          End of second year $\$170\,000$ End of second year $\$158\,950$
          End of third year $\$145\,000$ End of third year $\$135\,107.50$

2 a First year depreciation is $\$10\,000$ **b** First year depreciation is $\$16\,000$
          Second year depreciation is $\$10\,000$ Second year depreciation is $\$12\,800$
          Third year depreciation is $\$10\,000$ Third year depreciation is $\$10\,240$

3 a $\$23\,000$ **b** $\$16\,000$ **c** $\$9000$

Exercise 3E   PAGE 67

1 A: $\$182.738\,9691$ B: $\$182.74$ C: 12   D: $\$4000$ E: $\$0$
       F: 2   G: 9   H: monthly   I: 12

2 A = 5359,   B = 3894.90,   C = 512.83,   D = 564.11.

3 A = 1.015,   B = 1043.86,   C = 180.64.

4 $\$49\,700$, $\$49\,397$, $\$49\,090.97$, $\$48\,781.88$ (nearest cent).

5 a At the end of year five and immediately after his regular monthly repayment, Hisham still owes $\$15\,016.34$ on the loan.

   b Hisham would owe $\$12\,974.59$ at the end of the 5 years (after his repayment for the month).

6 a Pays off the loan at the end of the 70th month (and the 70th repayment is less than $\$250$).

   b Pays off the loan at the end of the 56th month (and the 56th repayment is less than $\$290$).

   c Pays off the loan at the end of the 48th month (and the 48th repayment is less than $\$290$).

7 The constant monthly repayment needs to be $\$443.21$, to the nearest cent.

8 The constant monthly repayment needs to be $\$717.35$, to the nearest cent.

9 The constant monthly repayment needs to be $\$1\,972.11$, to the nearest cent.

10 The loan would be paid off in exactly two years by paying $\$188.29$ per month for 24 months (actually a few cents more on the very last repayment to exactly pay off the outstanding amount).

    If instead the monthly interest rate was 1.2% per month the monthly repayments would need to be $\$192.81$ (now a few cents less on the very last repayment).

1 Loan of $514 214.53 + deposit of $17 000.
Hence, to the nearest $1000, Fran and Michael can afford a house costing $531 000.

2 Loan of $305 688.95 + deposit of $20 000.
Hence, to the nearest $1000, Peta and Peter can afford a house costing $326 000.

3 Loan of $441 419.57 + deposit of $35 000.
Hence, to the nearest $1000, Rania and Umah can afford a house costing $476 000.

4 Loan of $234 000.19 + deposit of $450 000.
Hence, to the nearest $1000, Hue and James can afford a house costing $684 000.

5 Loan of $364 609.29 + deposit of $30 000.
Hence, to the nearest $1000, Kirra can afford a house costing $395 000.

6 $3 575.13   (Note: Round *down* from $3 575.130 92. Final payment will be a little *over* $3575.13.)
$622 540 (nearest $10)   Note: $300 \times \$3575.13 - \$450 000$ gives $622 539.
Allowing for larger 300th payment gives $622 539.93.
Different rounding regime, e.g. on some online amortization calculators, gives $622 539.38.

7 $2 991.93   (Note: Round *up* from $2991.927 336. Final payment will be a little *under* $2991.93.)
$718 060 (nearest $10)   Note: $240 \times \$2991.93$ gives $718 063.20.
Allowing for reduced 240th payment gives $718 061.78.
Different rounding regime, e.g. on some online amortization calculators, gives $718 061.84.

Paying $100 more per month would pay off the loan approximately 16 and a half months earlier.

Miscellaneous exercise three   PAGE 72

1 a $11 754.62   **b** $11 887.58   **c** $11 918.77

2 A seasonal index of 1.08 for Autumn means that whatever it is that the index is referring to, e.g. sales numbers, people moving interstate, unemployment figures, etc, the measurements for Autumn tend to be 8% above the average for a season.

3

Season	Mon	Tue	Wed	Thur	Fri	Sat
Deseasonalised data	123	113	113	125	117	133

4 a $T_1 = 7$, $T_2 = 19$, $T_3 = 31$, $T_4 = 43$, $T_5 = 55$
 b $T_1 = 100$, $T_2 = 85$, $T_3 = 70$, $T_4 = 55$, $T_5 = 40$
 c $T_1 = 5000$, $T_2 = 6000$, $T_3 = 7200$, $T_4 = 8640$, $T_5 = 10 368$
 d $T_1 = 2000$, $T_2 = 200$, $T_3 = 20$, $T_4 = 2$, $T_5 = 0.2$
 e $T_1 = 4$, $T_2 = 11$, $T_3 = 25$, $T_4 = 53$, $T_5 = 109$
 f $T_1 = 200$, $T_2 = 296$, $T_3 = 440$, $T_4 = 656$, $T_5 = 980$

5 a D   **b** B, C   **c** A   **d** B, D   **e** B

6 a $48 620.70   **b** $45 810.01   **c** $41 460.41   **d** $32 257.88

7 a $16 483.73
 b 15.19 years, i.e. just over 15 years and 2 months.
 c 18 years.
 d i at least 11.3264%   **ii** at least 10.7777%

8

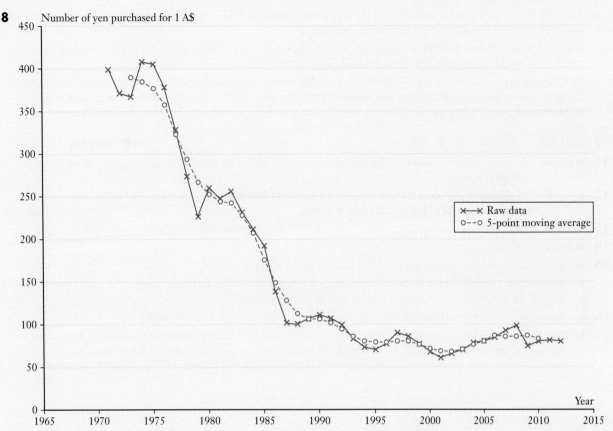

Number of yen purchased for 1 A$

Comments summarising the data, and about today's exchange rate, are not included here. Compare your comments to those of others in your class.

Exercise 4A PAGE 81

1 If $8000 (or more) is invested at 7.5% per annum then the interest earned will be $8000 × 0.075 i.e. $600 (or more). Hence the interest each year will equal (or exceed) the award amount. The balance in the account will then remain at (or exceed) $8000 thus allowing the award to continue 'in perpetuity'.

2 $1.02^4 = 1.082\,432\,16$. Thus 8% per annum, with quarterly compounding, increases an initial investment by 8.243 216%.
If the initial investment is $A we require 8.243 216% of $A to equal $75 000.
Thus $A = $909 839.0725, i.e. $909 840 rounded up to the next dollar.
The one off investment needs to be $909 840 (or more).
(Other approaches are possible, for example trial and adjustment.)

3 $243 199.29 (or more).

4 a Increase P. **b** An increase in R. **c** A decrease in A.

5 There will be $436 670 left in the account after the tenth withdrawal (nearest ten dollars).
Kelvin can withdraw $50 000 per year for 22 years. (Then, at end of the 23rd year, the withdrawal would be less than $50 000 and would close the account.)
Had the interest rate been 7.8% Kelvin would have been able to withdraw $50 000 per year for 45 years. (Then, at the end of the 46th year, the withdrawal would be less than $50 000 and would close the account.)

6 Julie will be able to withdraw $50 000 from the account for 26 years. (Then, at the end of the 27th year, the withdrawal would be less than $50 000 and would close the account.)

If instead Julie withdrew $45 000 per year she could do this for 34 years. (Then, at the end of the 35th year, the withdrawal would be less than $45 000 and would close the account.)

To maintain her standard of living it is likely that Julie will need an increasing amount to live on each year because of the likely rise in the cost of everything due to inflation. Hence she might be wiser to withdraw an amount which increases each year to allow for this inflation, even if it means having to withdraw less than the $50 000 in the early years.

7 $352 941   **8** $575 251   **9** $358 663   **10** $454 576

11 a and **b** Nineteen annual payments of $30 000 and then at the end of 20 years the final payment will be $5214.89 (= $4851.06 × 1.075).

12 a and **b** One hundred and seventy three monthly payments of $4000 and then at the end of 174th month the final payment will be $2534.41 (= $2524.31 × 1.004).

13 a and **b** $381 236

14 a and **b** $46 732

15 a Immediately after the third withdrawal the balance in the account will be $277 722.60.

b Immediately after the 14th withdrawal the balance is $26 395.75. Hence at the end of the 15th year, by withdrawing $28 507.41 (= $26 395.75 × 1.08) the balance will be reduced to zero.

16 a Immediately after the eighth withdrawal the balance in the account will be $149 506.05.

b Immediately after the 13th withdrawal the balance is $575.01 ($575.009). Hence could simply add this to the payment at the end of the 13th year and reduce the balance to zero or could continue for another year and at the end of the 14th year, by withdrawing $618.13 (= $575.009 × 1.075) the balance will be reduced to zero then.

Exercise 4B   PAGE 87

1 $259 162.18   **2** $567 873   **3** $333 114.12   **4** $565 537.24

5 Fifty seven quarterly payments of $15 000 followed by a final payment of $9924.23 which closes the account.

Computer spreadsheet or graphic calculator activity   PAGE 88

Answers not given here. Compare your answers with those of others in your class.

Miscellaneous exercise four   PAGE 89

1 $2448.90

2 The 855 is the gradient of the line of best fit. It informs us that over the ten years under consideration the number of people attending the music festival increased at approximately 855 people per year.

3 a 6.136%   **b** 6.168%   **c** 6.183%

4 $4267.45

5 a $4300   **b** 6.5%   **c** $6274.31   **d** 20.795%

6 The initial investment was $1600, at a fixed monthly interest rate of 1.5% with a fixed monthly deposit of $250.

7 a A: 0.7570, B: 0.8033, C: 0.7778, D: 0.7991, E: 160, F: 0.7625, G: 1.175

b

Calculation of seasonal indices

	2010	2011	2012	2013	2014	Seasonal index (3 dp)
1st 4 Months	1.1190	1.0233	1.1264	0.9953	1.0625	1.065
2nd 4 Months	0.7570	0.8033	0.7778	0.7991	0.7625	0.780
3rd 4 Months	1.1241	1.1733	1.0958	1.2056	1.1750	1.155

Seasonally adjusted sales figures, *S* (Nearest integer)

	2010	2011	2012	2013	2014
First 4 Months	415	288	276	200	160
Second 4 Months	383	309	260	219	156
Third 4 Months	384	305	248	223	163

c Using integer values for *S*: $S = -17.91t + 409.2$
(Using more accurate values for *S*: $S = -17.919t + 409.36$)

d For the first 4 months of 2015 predicted sales are   131   (nearest integer).
For the second 4 months of 2015 predicted sales are   82   (nearest integer).
For the third 4 months of 2015 predicted sales are   100   (nearest integer).

Exercise 5A PAGE 96

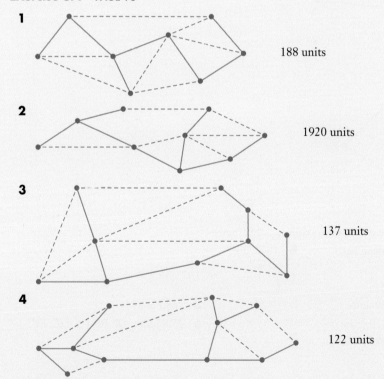

1   188 units

2   1920 units

3   137 units

4   122 units

5

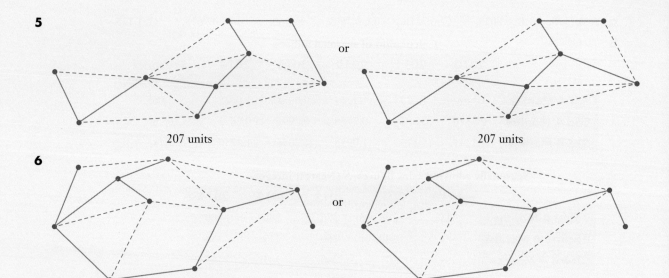

207 units   or   207 units

6

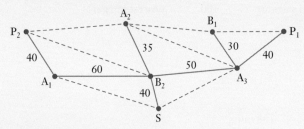

2250 units   or   2250 units

7 Replace AG, GF, FC, CB, FE, ED with new piping. Close off AF, AC, AB, CE, CD, BD. New piping has a total length of 325 m.

8 295 m

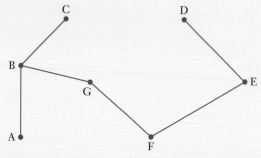

9 a

B
├─ C
A ─ B ─ G ─ F ─ E ─ D

b The total length of tunnelling that will be closed off to visitors is 445 metres.

10 a Graph showing roads and distances not given here.   **b** 1890 km

Exercise 5B   PAGE 102

1 76 units   **2** 129 units   **3** 1680 units
4 19.7 units   **5** 2520 units   **6** 25.8 units, AB, BF, CF, DE, EF
7 21.1 km   **8** $670
9 a 921 km
b Harlow - Raine,   Wayley - Gatley,   Wayley - Olber,   Raine - Lewis,   Raine - Fitch,   Fitch - Wayley
10 BF, CH, DF, EA, FE, GD, HG, IC, JC. 261 km.

Miscellaneous exercise five PAGE 105

1 a $N = 29.14n + 7668$

b The value of *a*, 29.14, means that linear modelling of the data suggests that during the time period the data covers the number of people employed full time is increasing at the rate of approximately 29 140 people per quarter.

c July 2011 has an *n* value of 7. The line of regression gives $N = 7872$.

Linear modelling gives a predicted employment figure for July 2011 of 7872 thousand.

January 2015 has an *n* value of 21. The line of regression gives $N = 8280$.

Linear modelling gives a predicted employment figure for January 2015 of 8280 thousand.

We would expect the July 2011 prediction, which involves interpolation, to be more reliable than the January 2015 prediction, which involves extrapolation. The latter involves predicting beyond known data points and events could happen in the meantime making such predictions unreliable.

d Viewing the data graphically, as shown on the right, suggests that linear modelling may be unwise. The overall trend seems more curved than straight. We could perhaps use two linear models but even then the apparent late downturn in employment suggests that extrapolating ahead of the given data based on a linear model could well be unwise.

The placement of the linear regression line shows why the predicted figure for July 2011 was below the real figures on either side of it. Hence using a linear model for this data even made an interpolated prediction doubtful.

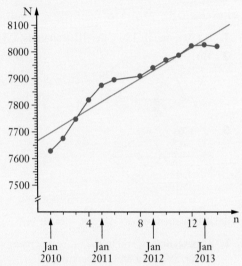

2 At 6% per annum, i.e. 0.5% per month, the initial $350 000 will earn interest of $1750 by the end of the first month. Hence a withdrawal of $1500 will leave more than $350 000 in the account for the next month. The account will continue to grow each month and the task set for the calculator, to find how many payments it takes to give a final balance of zero, is impossible. Hence the 'Error' response.

3 $a = 67,$ $b = 76,$ $c = 80,$ $d = 71.5,$ $e = 81,$ $f = 76,$ $g = 80,$ $h = 78,$
$i = 69,$ $j = 82,$ $k = 80,$ $l = 86,$ $m = 84,$ $n = 82,$ $o = 53.$

4 $3634.60

5 a At the end of year six and immediately after her regular repayment, Amy still owes $5636.26 on the loan.

b To pay off the loan in exactly six years Amy's annual repayments need to be $3264.44.

Exercise 6A PAGE 113

1 60 units   **2** 90 units   **3** 80 units   **4** 100 units   **5** 450 units
6 550 units   **7** 500 units   **8** 170 units   **9** 550 units   **10** 100 units

11 Max flow is 850 units.

One way it can be achieved is as shown.

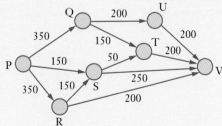

12 $v = 100,$ $w = 300,$ $x = 100,$ $y = 200,$ $z = 150.$

13 1800

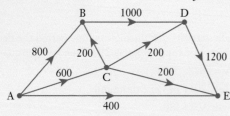

14 16 units

Upgrade AC to 8 units: Increase the maximum flow to 19.

Upgrading BD or FG: Will not alter the maximum flow.

Exercise 6B   PAGE 116

1 I: 150 units, II: 130 units, III: 160 units, IV: 140 units, V: 180 units.

2 450
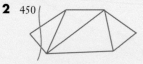

3
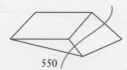
550

4 500
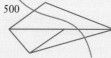

5 170
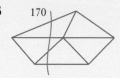

6

550

7
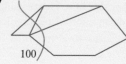
100

Miscellaneous exercise six   PAGE 117

1 As percentages, and to the nearest percent, the seasonal indices for the first, second, third and fourth quarters are 141%, 112%, 59% and 88% respectively.

Numbers for the first quarter tend to be 41% up on the quarterly average, for the second quarter 12% up on the quarterly average, for the third quarter 41% down on the quarterly average and for the fourth quarter 12% down on the quarterly average.

2 241 units

3 **a** $4026.14 needs to be repaid per month for the loan to be paid off in 25 years.
(Note: Rounding *up* from $4026.135417 so final payment will be a little *under* $4026.14.)

   b A total of $707840 (nearest $10) interest is paid during the 25 years.
Note: $4026.14 × 12 × 25 – $500000 gives $707842.
       Allowing for reduced 300th payment gives $707837.27.
       Different rounding regime, e.g. on some online amortization calculators, gives $707837.52.

4 **a** $12000   **b** 7.2%   **c** $18211.68

5 $14400 in first year, $12672 in second year, $11151.36 in third year.

6 1250 L/ minute.

(The maximum flow is 1250 L/min but the maximum flow diagram shown here is not the only possible way of achieving it.)

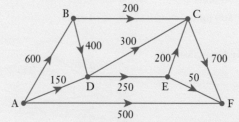

7 a Three-point moving averages are shown below. (Values are given to the nearest 100, as requested.)

n	3-pt MA (M)
1	–
2	1 837 500
3	1 828 600
4	1 794 000
5	1 830 100
6	1 850 500
7	1 894 900
8	1 930 000
9	1 938 800
10	1 937 900
11	1 923 600
12	1 944 700
13	1 964 700
14	2 010 700
15	2 037 400
16	2 076 800
17	2 126 800
18	–

b

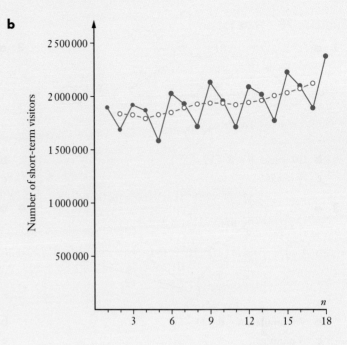

c $M = 18\,970n + 1\,753\,000$

d Interpretation of the number 18 970 in the equation $M = 18\,970n + 1\,753\,000$:
The number of short-term visitors arriving in Australia from January 2008 to December 2013 was, on average, increasing at a rate of approximately 19 000 people per four months.

Exercise 7A   PAGE 124

1 a 24 days **b** P T U W Y **c** 3

2 a 12 hours **b** A E F H **c** 12 hours (i.e. no change) **d** 1

3 a 17 days **b** A C H L N **c** 17 days (i.e. no change)

4 a 27 hours **b** P S V Z **c** No (cut by 1 hour, P W X Y becomes critical)
d P

5 a 30 days **b** A C G J **c** 14 days into the project. **d** 16 days into the project.
e 2 days **f** 4 days **g** 3 days **h** F

6 a 74 hours **b** P Q R S **c** 10 hours **d** 29 hours
e 19 hours **f** 57 hours **g** 10 hours

7 a 17 minutes, P T Y and S X Z **b** 2 minutes each.

8 a 44 days **b** B C G J **c** 14 days **d** 2 days
e 4 days

9 a 3 h 10 min **b** 12:10 p.m. **c** 11:05 a.m.

10 a 4:25 p.m. **b** 4:55 p.m. **c** 45 mins

1 a

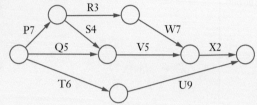

b 19 weeks, P – R – W – X

c 4 weeks

2 a

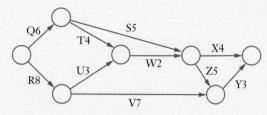

b 21 days, R – U – W – Z – Y

c 23 days

3 a

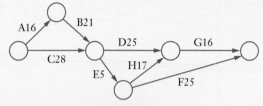

b 78 minutes, A – B – D – G

c 3 minutes

4 a

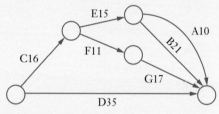

b 52 days, C – E – B

c 8

5 a

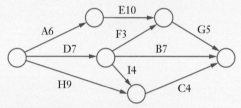

b 21 hours, A – E – G

c D and H

6

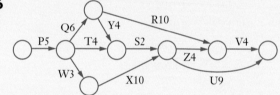

27 minutes, P – W – X – U

7 a

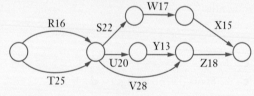

b 79 days, T – S – W – X

c 80 days

8

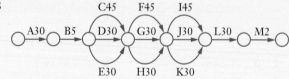

11:50 a.m.

9 a

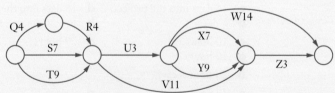

b 26 hours, T – U – W

c **i** Two hours over the allocated three.

   ii Three hours over the allocated three.

ISBN 9780170395069

10 a

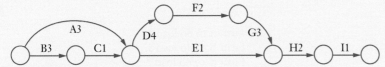

b 16 days, B – C – D – F – G – H - I

c 8 days

d Reduce by one day.

11

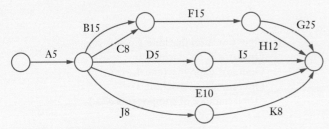

Minimum completion time 60 minutes.
Attempt to speed up:
Service vehicles getting in position, passengers on and off, security check.

12

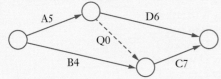

Critical path is A – Q – C
Minimum completion time is 12 hours.

13

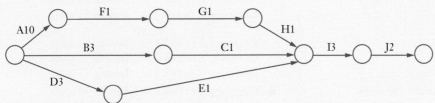

Minimum completion time is 18 weeks.

Miscellaneous exercise seven   PAGE 136

1 A seasonal index for January of 1.12 means that whatever it is that was used to determine the index, e.g. sales numbers, gross takings, number of people, prices etc., increases in January by 12% of the monthly average.

2 Five years after the commencement of the loan the amount owed is $243 862.70 (to the nearest cent).

3 Seasonally adjusted figures are 2208 for January and 2269 for February.

4 a A, C.   **b** B, D.   **c** A, C, D.   **d** C   **e** D

5 The maximum number of units that can be delivered is 750.

Increasing the capacity of AC to 250 units increases the max flow by 50 units (to 800 units).

6 Connections need to be: A to B, B to Base, Base to D, D to C. $460 000

7 The following pairs of lookout positions should have direct road connections constructed: 1 and 2, 1 and 5, 2 and 3, 3 and 4.

8 The greatest number of vehicles that could arrive at Z per hour, having come from A, is 2700.

1 Firm A to do load 2.
Firm B to do load 1.
(Total cost $240.)

2 Firm A to do load 1.
Firm B to do load 3.
Firm C to do load 2.
(Total cost $340.)

3 Firm A to do load 2.
Firm B to do load 4.
Firm C to do load 3.
Firm D to do load 1.
(Total cost $3100.)

4 Operator A on machine 1.
Operator B on machine 2.
(Total number of components 180.)

5 Operator A on machine 2.
Operator B on machine 1.
Operator C on machine 3.
(Total number of components 610.)

6 Operator A on machine 2.
Operator B on machine 1.
Operator C on machine 4.
Operator D on machine 3.
(Total number of components 375.)

1 Company A to task 2.
Company B to task 1.   Total cost 177 units.

2 Company A to task 2.
Company B to task 1.
Company C to task 3.   Total cost 300 units.

3 Company A to task 3.
Company B to task 1.
Company C to task 4.
Company D to task 2.   Total cost 43 units.

4 Company A to task 4.
Company B to task 5.
Company C to task 2.
Company D to task 6.
Company E to task 1.
Company F to task 3.   Total cost 135 units.

5 Company A to task 2.   Benefit 120 units.
Company B to task 1.   Benefit 240 units.

6 Company A to task 1.   Benefit 29 units.
Company B to task 3.   Benefit 28 units.
Company C to task 2.   Benefit 33 units.
or
Company A to task 3.   Benefit 31 units.
Company B to task 1.   Benefit 26 units.
Company C to task 2.   Benefit 33 units

7 Company A to task 2.   Benefit 240 units.
Company B to task 3.   Benefit 140 units.
Company C to task 4.   Benefit 610 units.
Company D to task 1.   Benefit 160 units.

8 Company A to task 2.   Benefit 40 units.
Company B to task 1.   Benefit 51 units.
Company C to task 3.   Benefit 42 units.

9 Company A to task 3.   Benefit 33 units.
Company B to task 4.   Benefit 43 units.
Company C to task 1.   Benefit 35 units.
Company D to task 2.   Benefit 45 units.

10 b Steps of algorithm not shown here. Optimal solution is as follows:
Adelaide rep oversees Sydney office.   (Return airfare $380.)
Brisbane rep oversees Darwin office.   (Return airfare $570.)
Hobart rep oversees Melbourne office.   (Return airfare $290.)
Perth rep oversees Broome office.   (Return airfare $720.)
(Total of return airfares is then $1960.)

11 a   Quick Courier Co. to job 1.   ($290)   or   Quick Courier Co. to job 4.   ($150)
Speedy Courier Co. to job 3.   ($260)   Speedy Courier Co. to job 3.   ($260)
Deliverit Couriers to job 2.   ($80)   Deliverit Couriers to job 2.   ($80)
Competitive Couriers to job 4.   ($140)   Competitive Couriers to job 1.   ($280)
Total cost of this minimum total cost solution is $770.   Total cost of this minimum total cost solution is $770.

b   If all four companies add $50 for job 4 the optimum allocation would not change, i.e. the two alternatives would still be as above (but the total cost would increase to $820).

c   If all four companies multiplied their prices for job 2 by 4 the optimum situation would change.

Quick Courier Co. would do job 1.
Speedy Courier Co. would still do job 3.
Deliverit Couriers would do job 4.
Competitive Couriers would do 2.
(Total cost of this minimum total cost solution now $970.)

d   Competitive couriers. (Total $730.)

12   Jack to area 1.   Jim to area 4.   Judy to area 3.
Nicci to area 5.   Ti to area 2.   Total sales 50.

13   Ashok to location 3.   Daniel to location 1.   Kirsten to location 2.   Shani to location 4.

14   Two possibilities:

A → 1 (40 km)   B → 5 (30 km)   C → 2 (30 km)   D → 3 (40 km)   E → 4 (40 km).

Or

A → 3 (30 km)   B → 5 (30 km)   C → 2 (30 km)   D → 1 (50 km)   E → 4 (40 km).

(Each possibility gives a total distance travelled of 180 km.)

15   Factory A to manufacture the Charger.
Factory B to manufacture the Exeon.
Factory C to manufacture the Bijou.
Factory D to manufacture the Devine.
Factory E to manufacture the Arcain.

Exercise 8C   PAGE 160

1   Assign Alex to task 3   (15 minutes)
Assign Ben to task 4   (17 minutes)
Assign Theo to task 1   (21 minutes)

Working simultaneously all three will have finished after 21 minutes.

Yes, if all working together they could finish three tasks in 20 minutes if Alex assigned to task 4 (20 minutes), Ben to task 1 (18 minutes) and Theo to task 3 (18 minutes). Working simultaneously the three tasks would be completed after 20 minutes.

2   Assign the digger from the Horton storage facility to Abbysix.   (25 km)
Assign the digger from the Peel storage facility to Crampon.   (18 km)
Assign the digger from Wimsley storage facility to Denton.   (31 km)
Total distance 74 km.

3   Assign taxi A to Tenielle, who will have to wait 8 minutes.
Assign taxi C to Zarani, who will have to wait 6 minutes.
Assign taxi D to Bill who will have to wait 8 minutes.

4   Assign person B to job 2.   ($220)
Assign person C to job 1.   ($120)
Assign person D to job 3.   ($160)
Assign person E to job 4.   ($150)
Total profit is $650.

Challenge PAGE 162

Optimum solution is

Worker A to task 9.	15 minutes.	
Worker B to task 6.	15 minutes.	
Worker C to task 4.	21 minutes.	
Worker D to task 7.	29 minutes.	
Worker E to task 3.	9 minutes.	
Worker F to task 10	15 minutes	
Worker G to task 1.	17 minutes.	
Worker H to task 2.	20 minutes.	
Worker I to task 5.	14 minutes.	
Worker J to task 8.	36 minutes.	Total 191 minutes.

Miscellaneous exercise eight PAGE 163

1

Compounding frequency	Nominal annual interest rate (%)	Effective annual interest rate (%)
Annual	10%	10%
Six monthly	10%	10.25%
Quarterly	10%	10.38% (2 dp)
Monthly	10%	10.47% (2 dp)
Weekly	10%	10.51% (2 dp)
Daily	10%	10.52% (2 dp)

2 **a** $553 382

   b $553 382

3 Correct answer is $662 642, to the nearest dollar.

4 Immediately after the fifteenth repayment has been made the amount owed is $4908.28 (to the nearest cent).

5 27 hours, B – E – J – M

6 **a** The greatest number of messages that can go from A to G in an hour is 850.

   b The greatest number of messages that can go from A to D in an hour is 650.

7 The maximum flow from A to X is 1200 units.

The extra capacity along EX increases the maximum flow from A to X by 150 units.

8 Fran should swim backstroke (54.2 seconds).
Marta should swim breaststroke (48.2 seconds).
Julie should swim butterfly (43.6 seconds).
Micah should swim freestyle (42.4 seconds).
Total time 188.4 seconds.

9

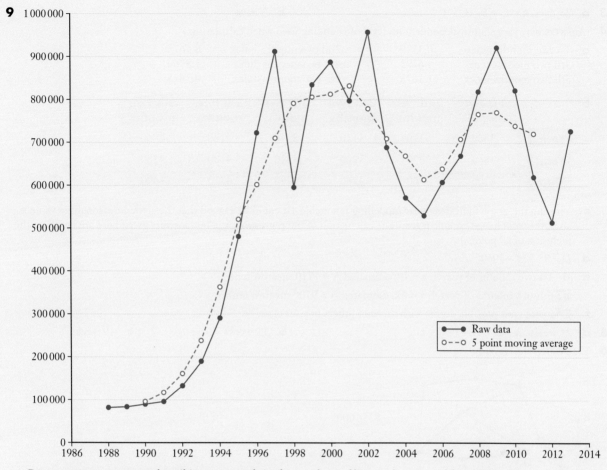

Compare your comments describing any trends in the numbers of live cattle exported from Australia per year over the period of time from 1988 to 2013 with those of others in your class.

10
Vehicle A to location 3	7 minutes.
Vehicle B to location 2	9 minutes.
Vehicle C to location 1	6 minutes.
Vehicle D to location 5	9 minutes.
Vehicle E to location 4	10 minutes.

11 a 6850

b
1st Quarter	4932
2nd Quarter	6028
3rd Quarter	7261
4th Quarter	9179

12 The total length of the minimum spanning tree for the given network is 100 units.

The minimum span is shown on the right.

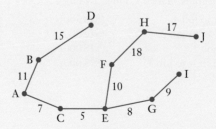

13 a 25 days, A – C – F – G – J **b** 3 days

14 (Answers may vary slightly dependent on level of rounding used when calculating.)

a
1st two months index 0.3579	2nd two months index 0.5675
3rd two months index 1.6293	4th two months index 1.8760
5th two months index 1.1878	6th two months index 0.3815

b

	1st two months	2nd two months	3rd two months	4th two months	5th two months	6th two months
Year 1	3460	3190	3120	3370	3180	2760
Year 2	3860	3870	3940	3810	4100	4280
Year 3	4200	4560	4580	4380	4310	4690

c Viewing the graph indicates linear modelling is suitable. (It could be argued that the sixth deseasonalised value is an outlier and hence exclude it from the calculation of the regression line but answers given here are based on the inclusion of all points.)

d $D = 95.52n + 2960$

e **i** 1st two months of year 4, $n = 19$, estimated $N = 1710$ (nearest ten).

ii 4th two months of year 4, $n = 22$, estimated $N = 9500$ (nearest ten).

f Compare your possible events with those of others in your class.

15 a 21 weeks, A – D – H – I – J – L **b** 11 weeks **c** 19 weeks

16 $256 000

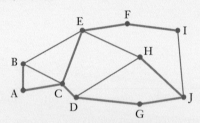

$256 000

Direct connection from H to I is worth doing if it can be done for less than $38 000 as it could then replace EC in the minimum span.

17 Either assign

Machine A to product 3	(3300 units)
Machine B to product 4	(4100 units)
Machine C to product 1	(4000 units).

or assign

Machine A to product 4	(3300 units)
Machine B to product 3	(4100 units)
Machine C to product 1	(4000 units).

(Either way gives the total number of units produced as 11 400.)

18 a 1155 m **b** Goods In - Dock 1, Customs - Goods Out, Customs - Waiting, Customs - Dock 3, Dock 1 - Dock 2, Dock 2 - Customs.

19 Allocate DG Transport to job 4. ($16 500)
Allocate Haulage and Co. to job 2. ($12 800)
Allocate JA Movers to job 1. ($18 000)
Allocate Move it and Shift it to job 3. ($25 000)
Total cost $72 300.

ISBN 9780170395069

INDEX

ISBN 9780170395069

S

scattergraphs xiv
seasonal effects 9, 20
   quantifying 33–5
seasonal indices 33–5
seasonally adjusted data 39–40
   making predictions 40–3
simple interest x, 53–4, 56
simultaneous equations ix
sink (maximum flow) 109, 110
slack 122, 123
smoothing data 20–2, 26
solving equations ix
source (maximum flow) 109, 110
spanning trees 93–5, 99–101
spreadsheets
   compound interest 60–2, 66
   geometric sequences xvii
   moving averages 27
square matrix (allocation problems) 157–9
statistical investigation process xviii, 4
straight line graphs viii
superannuation 77–9
   accumulation phase 77
   pension phase 77

T

technology use xvi–xviii
3 point moving average 20–1
time series data 3–6, 14–15
   general smoothing 26
   making predictions 6–10
   seasonal variation 9
total float 123
trees xii
   spanning 93
trend line xiv
   making underlying trends more apparent 20

U

underlying trends 20
unit cost method 64

V

vertices xi

W

whole life annuities 79